STRESS

and

VISION

Stress and Vision

by Elliott B. Forrest, O.D.

Library of Congress Catalog Card Number: 88-60463
ISBN: O-943599-00-8

Printed in the United States of America

Published by Optometric Extension Program Foundation, Inc.
2912 South Daimler Street
Santa Ana, CA 92705-5811

Managing Editors: Frances Conkey
Sally Marshall Corngold

First Edition, 1988

Elliot B. Forrest's expertise and excellence as a scientific writer, educator, lecturer and practitioner are legend and his contributions to the optometric profession are well documented.

The Optometric Extension Program Foundation is pleased to publish this book as a tribute to his memory and in recognition of his many talents. The concepts and views presented herein are the sole product of the author and do not necessarily represent the opinions or philosophy of the publisher.

March 1, 1988

About The Author

Elliott B. Forrest, O.D., was an Associate Clinical Professor of Optometry at the State College of Optometry, State University of New York where he was on the teaching staff of the Vision Training Department and Co-Director of the Infants' Vision Clinic of the University Optometric Center for 15 years.

Dr. Forrest lectured and published extensively. Major publications include a 15-part series entitled "A Modern Philosophy of Vision" (Optometric Weekly, 1959-1960); a 24-part series entitled "The Psychobiology of Visual Behavior" (Optometric Extension Program Postgraduate continuing education courses, 1968-1970); and a text, "Visual Imagery: An Optometric Approach," published by the Optometric Extension Program Foundation (1981). He is also noted for his work on visual stress, visualization, visual information processing, written language deficits and functional astigmatism.

Dr. Forrest graduated from the Northern Illinois College of Optometry in 1949. He was a member of the Board of Examiners of the College of Optometrists in Vision Development and Eastern Regional Chairman and Chairman of the Committee on Senior Student and Recent Graduate Research of the Optometric Extension Program Foundation. He was a Fellow of the American Academy of Optometry and the College of Optometrists in Vision Development and was elected to membership in the National Academies of Practice as a Distinguished Practitioner in Optometry.

Dr. Forrest, who conducted a private practice in West Hempstead, New York, was the recipient of many prestigious awards. These include the Samuel M. Lyons Award (1972) and the Distinguished Achievement Award (1978) from the New York State Optometric Association: the A. M. Skeffington Award for excellence in optometric writing (1980) and the President's Award (1982) from the College of Optometrists in Vision Development; the Distinguished Service Award (1984) from the New York Academy of Optometry; and the Distinguished Merit Award (1984) from the Nassau County Optometric Society.

Dedication

- Libby, without whose love and devotion there would be nothing to dedicate.
- the late A. M. Skeffington, O.D., Education Director and major theorist for the Optometric Extension Program Foundation, whose personal encouragement to search, to learn and to share started my own adventure in behavioral optometry.
- those many optometric colleagues and friends whose support and affection I have cherished and whose own dedication to the profession has been an inspiration.
- the future of behavioral optometry.

Acknowledgements

Grateful acknowledgment is given to the following for permission to reproduce or adapt illustrations from:

- Hans Selye, The Stress of Life, rev. ed. (New York: McGraw-Hill Book Co., 1978). Reproduced by permission of the publisher (Figure 2).

- Stress and Distress in Response to Psychosocial Stimuli, Lennart Levi, ed. (New York: Pergamon Press, 1972). Reproduced by permission of the publisher (Figures 3 and 19).

- G.S. Everly and R. Rosenfeld, The Nature and Treatment of the Stress Response (New York: Plenum Press, 1981). Reproduced by permission of the authors and publisher (Figure 4).

- Hans Selye, "Stress, Cancer and the Mind," in Cancer, Stress and Death, ed. J. Tache, H. Selye and S.B. Day, (New York: Plenum Press, 1979). Reproduced by permission of the editors and publisher (Figure 20).

The author further wishes to acknowledge and convey his appreciation to:

- Marion Eaton, former Biomedical Illustrator and Coordinator of the Center for Biocommunications, State College of Optometry, State University of New York for her talents and effort in preparing the majority of illustrations in this book.

- Steven Goodman, Chief Medical Photographer, Center for Biocommunications, State College of Optometry, State University of New York for the prepublication photographic preparation of the illustrations.

Table of Contents

List of Figures

List of Tables

Forward

Dr. Elliott Forrest was unique in his ability to integrate and synthesize knowledge from many disciplines, to see the clinical implications for optometry, and to develop innovative models and clinical care regimens. He was a clinician and an academic, caring for patients both in his private practice and at the institutions which he served, while steadily seeking to expand his knowledge and understanding of the visual system and its relationship to broader aspects of behavior.

During the 1960s and 70s, Dr. Forrest served on the staff of the Developmental Vision Clinic of the Optometric Center of New York, reflecting his early interest in the relationship between vision, development and learning. Later he was appointed to the faculty of the College of Optometry of the State University of New York, where he taught in the Vision Training Department and served as Chief of the Infants' Vision Clinic. He was greatly respected as a teacher, and was voted an Excellence in Teaching Award by his students.

Dr. Forrest's contributions were highly regarded by his peers as well. The College of Optometrists in Vision Development presented him with its highest honors, the A. M. Skeffington Award for contributions to optometric writing and the G. N. Getman Award for contributions to developmental optometry, as well as a special President's Award. The New York Academy of Optometry, the New York State Optometric Association and the Nassau County Optometric Association have each honored him with distinguished awards on numerous occasions.

Dr. Forrest was a major contributor to the emerging discipline of behavioral optometry. In his lectures, his writing and his teaching, he clarified relationships, shared insights, synthesized knowledge from other disciplines, and ultimately developed an integrated model of the relationship between vision and behavior.

Behavioral optometry is concerned with the effect of visual function on behavior, as well as the influence which organismic behavior exerts upon vision. Visual function impacts upon behavior, posture, achievement and performance; information- processing style and environmental interacts in turn influence both the structure and function of the visual system.

In recent years, research has documented that vision is influenced by environmental factors. Early experience influences brain function and structure. Myopia is more prevalent in humans and animals exposed to constricted environments and nearpoint activity. Adaptive mechanisms in the vergence system operate to restore preexisting heterophoria when such heterophoria is neutralized with prism. Indeed, function determines structure just as structure determines function; the relationship is mutually interactive. Likewise the interaction between vision and the environment: we use vision to derive information about the environment, and our visual system can be altered as a product of these environmental interactions.

The relationship between man and his environment, between structure and function, and between visual function, information-processing, academic achievement and human potential pervades Elliott Forrest's writing and thinking. He dealt with these themes in lectures and in numerous papers on visual information- processing, visual imagery, astigmatism, stress and infants' vision, as well as in three major texts: "A Modern Philosophy of Vision," "The Psychobiology of Visual Behavior" and "Visual Imagery: An Optometric Approach."

In "Stress and Vision," Dr. Forrest's posthumously published final text, he raises questions of the utmost significance:

> What is stress?
> How does stress affect the visual system?
> Why are these effects more pronounced in some individuals than others?
> How does visual information-processing style influence stress reactivity?
> Does stress reside in the nearpoint task, or in the individual? Are we the victims of our environment, or do we create our own stress? And what can we do about it?

"Stress and Vision" presents Elliott Forrest's most advanced and most controversial thinking. His psycho-behavioral approach moves beyond the Skeffington view that nearpoint stress is primarily task-related. He suggests that mental states are the primary factors which underlie functional vision disorders. Although vision problems may be triggered by environmental interaction, Forrest concludes that the fundamental causes are rooted in our attitudes and belief systems. Dr. Forrest relates the nearpoint stress concept to general stress and to the behaviors which underlie it. He reviews the physiology of stress and its systemic and visual effects, and discusses the psychological factors which trigger the physiological stress response. He suggests that individuals who are intense and analytical, who engage diligently in a task with maximum visual concentration and strong desire for achievement, attempting to get the work done in as short a time as possible, who emphasize focal vision with decreased peripheral awareness, exhibit a greater stress response. Such individuals are thus more prone to stress-induced visual and systemic disorders. Dr. Forrest suggests not only that psychological factors predominate in determining who develops nearpoint stress-induced vision problems, but even further, that the specific adaptive outcomes manifest in a particular individual reflect his mental states, attitudes, beliefs and information-processing style. At some level of consciousness, we choose the disorders which afflict us. He offers the strongest case yet presented for this notion, synthesizing research and theory in psychology and physiology to develop a psycho-behavioral model of vision. This model has major implications in clinical care. It suggests that the role of the clinician is not simply to remediate the ocular and oculomotor deficits which are present, but to influence and modify the behaviors and attitudes which foster these deficits. In conjunction with traditional approaches, optometrists can deal with the mental states which underlie the visual problem, to guide patients in more appropriate directions. Dr. Forrest recognizes that changing belief systems is difficult, since we are typically unaware of our core beliefs and attitudes. He makes numerous suggestions for reducing stress and for changing attitudes. "Stress and Vision" presents not simply a philosophy of vision care, but a philosophy of life itself. It is Dr. Forrest's most significant contribution, and constitutes a landmark in the literature of behavioral optometry.

Martin H. Birnbaum, O.D.
New York, July 1987

Chapter 1

Introduction: A Personal Adventure

This book is about the stress response, a subject that I have been close to since my early days in optometry. In fact, one can say that this book has taken thirty-five years to write. It is a subject that I first formally began to research and write about in the mid-1950s and then not again until 1980, almost twenty-five years later. The reason for this prolonged period of silence is a tale worth telling.

It all started with my first exposure as an optometry student to the functional concept of vision as formulated and taught by A. M. Skeffington, O.D. This took place after having been dutifully indoctrinated in such obligatory courses as anatomy, physiology and geometric optics. The cornerstone of Dr. Skeffington's functional approach was the concept of visual stress.

Skeffington (1950f) viewed visual stress as resulting from "...the biologically unacceptable, socially compulsive, visually near-centered task which provokes an avoidance reaction...that becomes a drive to center nearer in visual space." Refractive problems or ocular defects were seen specifically as being the end result of adaptation to visual stress.

This concept had some profound implications. It implied, for example, that vision is not a static, rigid, mechanical optical system but an active, dynamic process that is responsive to the interaction of the organism with the environment. It also implied that certain types of interactions, such as prolonged near-vision tasks, can trigger a specific type of adverse response, an accommodatively induced overconvergence tendency. In addition, if this adverse response persists over a prolonged period of time, certain types of adaptive reactions, at both near and far distances, become predictable.

Further extensions of this model led to the ability to extrapolate from a set of examination findings information as to whether an individual has been exposed to persistent visual stress, how that stress had affected him or how he had coped with it, approximately how long this coping may have gone on, and how performance might have been affected.

In spite of the fact that this concept of visual stress with its profound insights had proven to be a viable operating framework for clinical diagnosis and therapy, it lacked a solid physiological base to explain what was occurring. It was mainly derived from clinical observations. Somehow, I personally needed to feel that in some way this view could be reconciled with known physiological research. This feeling was no doubt fostered by the many years I was involved in studying the biological sciences. I fully realized that the reason many behavioral researchers avoided physiological explanations and preferred to think of the organism or the brain as a "black box" was that it was more fruitful to observe the response to a stimulus and then infer what must be going on than to rely on limited existing neurological or physiological knowledge alone. Still, I felt that there should be some way of matching, however loosely, one's clinical observations with whatever was currently known about physiology. It was also very apparent that there were many gaps in functional philosophy that were being glossed over a little too easily.

Simply to clear things up in my own mind, I undertook the project of writing an essay about some of the key issues involved in functional optometry and counterpointing them, wherever possible, with available information from physiological research. Since we were involved with the same human organism, different approaches should have a great deal in common, as long as one is open to consider commonalities. Too often, unfortunately, we tend to stress differences in viewpoints rather than what is common among them.

This attempt at reconciliation was eventually published in the Optometric Weekly in late 1959 and early 1960 as a fifteen-part series entitled "A Modern Philosophy of Vision" (Forrest, 1959-60a). The key element in this series was a section called "The Story of Stress." It was an attempt to relate the work of Hans Selye, M.D., with the concepts of A. M. Skeffington, O.D.

As some degree of personal expertise was developed in this area, I began to realize not only how little I knew but how little was actually known. While so many appeared to be content that stress theory could account for the functional origin of different refractive and performance conditions, I became increasingly disturbed by the many exceptions, inconsistencies and incongruities that I could now see so clearly.

Many questions repeatedly kept running through my mind. For example: What made a near-centered task "biologically unacceptable?" Why the differentiation between "socially compulsive" and "non-socially compulsive" tasks? What did Skeffington really mean by the term "avoidance reaction" and, more importantly, how did it result in a drive to "center nearer in visual space?" Were reading and close work, per se, the major triggers for what was called the nearpoint stress response or were other factors involved that were equally, if not more, important? Why did so many avid readers and good achievers not develop ocular defects such as myopia? Why did some myopes seem to get worse with a full minus prescription while others appeared to remain at the same level or even improve? If myopia and esophoria were indicative of being the result of a preference for "central" processing as many felt, why were not all myopes esophores? If 0.75 diopters of hyperopia was considered to be a desirable state according to functional thinking, why did so many low hyperopes have visual discomfort of some sort (asthenopia)? Why did some high hyperopes, who were avid readers, appear to have little or no visual asthenopia? What type of behavioral approach to the world did the high hyperope represent? Was emmetropia an entity in and of itself, representing something positive and definitive, or was it simply the exhaustion of a "buffer," a minor point in a broad continuum? What did all the varieties of astigmatism really represent functionally? If embeddedness represented the result of a particular way of operating over time, why did so many individuals with long- standing problems, who did not avoid close work, often demonstrate a low degree of embeddedness, while others with short-term problems, such as preschoolers or new readers, often demonstrate deep-seated embeddedness? What made the accommodative-convergence/accommodation (AC/A) ratio so resistant to change? What factors determined who became a B1-, B2- or C- type case? Could one change from one to the other? If so, what elements would precipitate it? Also, why were we seeing fewer and fewer B1 cases? What, really, was a phoria? Did it represent a muscle pull? Was it a problem in neural innervation? Was it a measure of inappropriate spatial localization? Could a phoria be all three at the same time? Or, was something else involved that perhaps included all three while superseding them? In terms of body posture, was a change in phoria the result of a postural change, the cause of a postural change or could both be the result of the action of a third element in many instances?

These were by no means the only questions. Yet, even today, I wonder how many people have more than superficial answers to most of these questions. In any case, at the time, I became quite sobered by what I did not know. More than that, there were few experts in the area who were responsive to a true dialogue on the subject. I found that those with a physiological bias did not seem to care to discuss functional approaches and most "functionalists" seemed to have a total aversion to being involved with physiological factors, general, neural or ocular.

Therefore, because of my unresolved questions about such a core issue, I decided to leave the subject of stress alone, at least until I could begin to understand it better and fill in some of the gaps and unknowns that I felt were quite significant. This adventure took me into many different areas: perception, problem-solving, performance, movement, sensory-motor matching, feedback monitoring, the psychology and physiology of visual behavior, the history of written language, visual information processing, internal representation, operational style, visual imagery and astigmatism, among others (Forrest 1967,1968-70,1972,1976a,1978, 1981a,1981c).

After close to twenty years of involvement with these many diverse areas and integrating what I learned from them, I found that I had developed a broader and more encompassing perspective of visual and human behavior, what I began to call a "psycho- behavioral" approach. Almost everything I learned along the way, optometric and non-optometric, seemed to fit in with everything else, including the stress complex.

Most important of all was the realization that I still did not have the answers to all my questions. In fact, the more "knowledge" I accumulated, the more questions and enigmas seemed to appear. The difference was that it was no longer intimidating. Instead of feeling a lack, I now had a liberating feeling that there would never stop being enigmas to perplex and worlds to conquer. So, after all those many years, I had to agree with T. S. Eliot ([1909-62] 1963) when he wrote:

And the end of all our exploring
Will be to arrive where we started
And know the place for the first time.
("Little Gidding")

Having come full circle, I re-entered the arena by preparing a short philosophical position paper on stress for the 1980 Invitational Skeffington Symposium, Washington, D.C., which was later published in the Journal of the American Optometric Association (Forrest 1980a) under the title, "Stress: A Re- definition." This material was then incorporated into a more comprehensive presentation on the subject that was given to the thirteenth annual meeting of the College of Optometrists in Vision Development (COVD) in November 1983. It was that presentation that led to this text.

This book will involve an exploration into the subject of stress and its many ramifications. It will highlight possibilities and probabilities. Though many ideas will be presented regarding the visual aspects of the stress response, the major emphasis will be on understanding the stress response in general. The reason for this is that the real answer to what is "biologically unacceptable," as Skeffington put it, about a near-vision task rests with a better knowledge of the psychophysiology of general stress. Toward that end, almost every area that will be touched upon, even when it might appear far afield, is intimately related to what occurs in the visual stress response.

It is quite obvious that the subject of stress has become extremely popular today in all quarters. There are an ever increasing number of books, articles and television presentations on some aspect of the subject. There are also an increasing number of individuals and disciplines that have become involved in the areas of stress and stress reduction. Everyone talks about stress: physicians, lawyers, clergymen, teachers, engineers, optometrists, dentists, executives and laborers. It is the "in" thing. People jog, diet, meditate, take vitamin supplements and go for various forms of therapy, all in the name of reducing stress. Are these approaches meaningful or has "stress reduction'" become another fad? More importantly, what is its relevance in terms of optometric diagnosis and therapy? These are the factors that this book will attempt to probe.

Since this book is written by an optometrist, it will touch on issues involving stress and the visual process that are of specific interest to the author and to behaviorally-oriented optometrists. Because of its scope, however, it will also touch on issues that are of equal importance to anyone interested in the physiology, psychology and philosophy of stress, especially newer views of the subject and the highly relevant conclusions to which they lead.

This is not a book that attempts to attract by being sensational or guaranteeing maximal self-improvement in minimal time. Its goal is to survey the subject and understand it better.

To further orient the reader, this book has a strong psychological or psycho-behavioral bias. It also uses an occasional philosophical frame of reference in order to examine the basis for common behavioral observations. Finally, though the author is an optometrist and is biased toward the dominance of vision and visual information processing in the intact organism, the approach in this book treats vision as a process that subserves consciousness, awareness and attention. It also recognizes that for the most part, stress does not come from external factors as much as it arises from within ourselves. In essence, it is us.

With this as an introduction, let us begin the search for a path through the wilderness known as "stress."

Chapter 2

The Origin of the Stress Concept

The stress response is not a new phenomenon. It has been around as long as living things have been around. It is concomitant with life. It is part of human nature. It is the way life reacts to living.

Nor has this phenomenon gone unnoticed over the ages despite the tendency to think of ourselves as being much more knowledgeable and sophisticated than those of generations past. The words may not be the same from generation to generation but it has always been observed that people respond to certain situations differently than to others and that different people often have different yet predictable reactions to the same event. Whether one uses the terms stress, pressure, anxiety, depression, strain, startle, upset, irritability, nervousness, overreaction, panic, temper, fatigue, irritation, compulsion or their equivalents, it is obvious that physical and emotional changes of some sort are related to anticipation, surprise, fear, fright, pain, anger, rage, effort, boredom, interest and joy. It may not have been scientifically labeled and quantified or experimentally validated but an awareness and even a description of these states can be found in the most ancient recorded literature.

The modern, more scientific view of stress, however, is based on the pioneering work of only a few individuals. The first is Claude Bernard (1859, [1865]1957, 1878). Although Bernard was not involved with the stress concept as we know it today, he undoubtedly deserves the honor of being considered the originator and catalyst of the type of thinking that led to modern stress theory. Bernard was a brilliant French physician, physiologist, researcher and philosopher who noted more than one hundred years ago that one of the most characteristic features of living creatures is their ability to maintain an internal equilibrium in the face of changes in the external environment. To Bernard, the preservation of a constant internal environment was the prime objective of all vital organisms. He considered disease to be due primarily to a breakdown in the body's ability to maintain this internal equilibrium or "constancy."

This dynamic approach to how an organism maintained its stability and coped in the midst of all kinds of disruptive external forces strongly influenced Bernard's contemporaries and the path of future thinking and research.

Using Bernard's concept as a base, C. Judson Herrick, around 1924, contributed the notion of "biological regulation." Biological regulation, to Herrick (1924), was not simply the ability of the organism to adjust to specific externally induced situations such as temperature changes or shifts in pH. Herrick viewed it broadly, as representing a "continuous readjustment of the organism to the flux of surrounding conditions...," to life itself, so to speak.

At about the same time, Walter B. Cannon (1929,1932), a Harvard physiologist, honed in on two aspects of this process. The first was the process, as Bernard described it, of readjusting to normally changing conditions so that an internal constancy or "steady state" could be maintained. This Cannon now called "homeostasis" from the Greek homoios (similar) and statis (position), meaning the ability to maintain the same state or to continue the same way.

The second was the process by which the body mobilizes itself for action in the face of a perceived threat. This Cannon called the "fight or flight" response. "Fight or flight," in this sense did not represent the final action taken but the preparation on the part of the organism for either alternative, attack or retreat.

While Bernard, Herrick and Cannon set the stage for understanding the organism's responsiveness and adaptability to circumstances that might tend to throw it off balance, it was Hans Selye (1976), a physician, who, without doubt, made the greatest impact on modern stress theory as we now know it and has become its uncontested father. His contribution has been so seminal that his ideas, whether one agrees with all of them or not, have become the baseline to which all current work is matched.

The Selye Story

Selye's story starts in 1926 when he was a sophomore in medical school at the University of Prague. It was about the same time that Herrick and Cannon were formulating and publishing their concepts. As a second year medical student, Selye noted that patients suffering from many diverse diseases ranging from terminal cancer, measles, arthritis right down to the common cold, all exhibited a series of similar signs and symptoms such as loss of appetite, loss of muscular strength, loss of ambition and loss of weight. Even their facial expressions betrayed the fact that they were sick. In effect, Selye (1970) uncovered a new syndrome, the syndrome of "just being sick."

At the time, Selye's superiors did not quite appreciate this new view. They taught differential diagnosis which, even as it is taught today, stresses the importance of differences that exist among things that appear grossly similar. Selye, on the other hand, stressed that the commonality behind the diversity, the syndrome that appeared to be superimposed over each of the individual diseases that his patients had, was equally if not more significant.

Ten years later, in 1936, the problem reappeared in a completely different manner. As an assistant professor of biochemistry at McGill University, Montreal, Selye was attempting to isolate new types of hormones from the extracts of cattle ovaries. In the course of his work, he injected various ovarian extracts into his laboratory animals. Surprisingly, all of the extracts, regardless of how they were prepared or whether they were pure or not, produced a series of very specific effects that were the same in each case.

Selye was quite intrigued by this result. In order to determine why this was happening, he decided to inject different non-ovarian extracts into his laboratory animals. He used extracts from placentas, kidneys and spleens. He also used adrenalin, insulin and formalin. He exposed animals to x-rays, cold and heat and even put them through forced exercises and prolonged immobilization. Regardless of what he did, in each and every instance, the very same physiological effects occurred, a triad consisting of:

1. enlargement of the adrenal cortex
2. deep, bleeding gastrointestinal ulcers
3. atrophy (involution or shrinkage) within the thymicolymphatic system (the thymus, spleen, lymph nodes and other lymphatic structures), accompanied by a diminished lymphocyte and eosinophil count.

This syndrome was so consistent that Selye soon realized that it could be used as an index of the way an organism appeared to react to insult, injury or just to certain circumstances. This index, however, needed a name. Selye, for better or worse, chose the name, "stress." Unfortunately, this choice created a problem in semantics.

The Semantic Problem

In physics,"stress" represents the force that tends to deform a body. The resulting deformation is called "strain." In other words, stress is the effector while strain is the effect.

Similarly, in general speech, "stress" usually refers to the agent that acts on the individual. We say that we are under stress, that a job is stressful. We often refer to stress as something that puts "pressure" on us.

Cannon (1935), in discussing homeostasis, also considered the external factors as stresses and the bodily changes that occur in maintaining the functional equilibrium as strain.

Selye, on the other hand, uses the term quite differently. To Selye, stress is the response. The many factors that produce the response are called stressors.

To complicate this even further, some stress researchers use the term stress to represent the causative factors. Others go along with Selye's usage and use the term to represent the bodily response to stressor agents. Others call the stressor agents "stress stimuli" and the bodily response the "stress reaction." Some also use the term stress to represent emotional anxiety and psychological pressure. Still others use the term stress to represent all of these factors or any combination of them even in the same sentence so that, as one researcher has put it, "Stress, in addition to being itself, and the result of itself, is also the cause of itself."

Selye apologized for this predicament, admitting that had his English been better when he was looking for a label for his syndrome, he would have called it the "strain" reaction. However, by the time he was able to distinguish properly between the words "stress" and "strain," the use of the term stress was so widely accepted, being called le stress in French, el stress in Spanish, lo stress in Italian and der stress in German, that it became impossible to change. It took on a life of its own. Selye understood the confusion that he inadvertently created and acknowledged that it was difficult for some not to use "stress" as the causative agent since it had been used this way for so long. In any case, the important factor is for one to understand clearly how the term is used whenever it is used. The general thrust of its use in this book will be according to Selye's definition.

The Concept of "Non-Specificity"

Selye (1974b) defines stress as "the non-specific response of the body to any demand made upon it." This definition emphasizes three important factors: stress is a reaction or response to something; the response can be produced by any agent, event or circumstance acting on the body and; the response is non- specific.

The term non-specific refers primarily to the fact that the same or similar response can be caused by all kinds of agents. This means that regardless of the stressor, what it is, how intense it is, or how unique it is, whether it is a bee sting, an infection, a bruise or a school examination, some part of the body's total response is stereotyped and exhibits identical neurochemical changes in each case. Actually, defining stress as a non-specific response can be somewhat misleading since non-specificity, in Selye's sense, relates to the precipitating causes and not the response. The response, which comprises a syndrome, is quite stereotyped and specific. What induces it is non-specific or variable in nature (Selye 1976). Selye's definition could perhaps be made more appropriate if it defined stress as "the non- specifically induced response of the body to any demand made upon it."

Every stressor, therefore, according to Selye, can be thought of as eliciting two effects (see Figure 1). First is a specific effect which is quite individual and distinctive and which aids in identifying the causative agent through differential diagnosis. It is through such specific effects that a bee sting, measles, a common cold and the fear of visiting a dentist can be differentiated from each other. The second is a non-specific (stressor) effect, which is a syndrome that is common to all stressor agents. It is this non-specifically induced effect that comprises what Selye calls the "stress response."

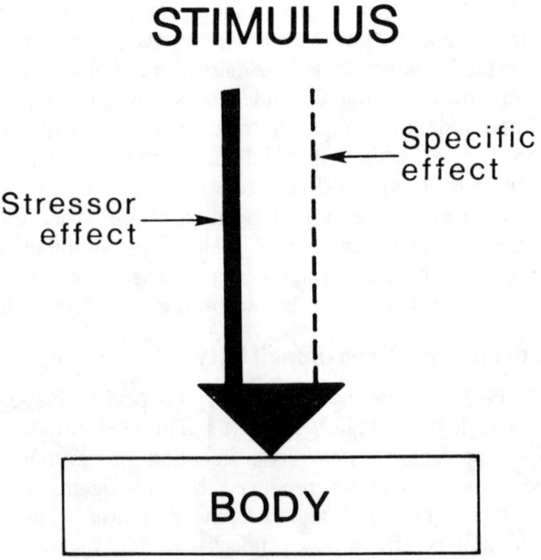

Fig. 1 The specific and non-specific aspects of the stress response

Since stress or the stress response results from any and all demands placed on the body, stress becomes essentially unavoidable. Even doing nothing at all is stressful. In fact, according to Selye, complete freedom from stress means only one thing--death.

Stress then, in Selye's terms, represents the end result of all one's interactions with one's environment. It maybe induced through changes in physical or physiological state by general movement, exercise, temperature shifts, humidity changes, noise, pollution, trauma, disease or diet. It can be activated by social interactions such as marriage, divorce, death, economic changes and all kinds of personal relationships. And, it can be triggered psychologically through one's attitudes, desires and emotions.

The concept of non-specificity, therefore, implies that though the stress response has a particular form, it has no particular cause. Its origins are ubiquitous.

The specific and non-specific responses, however, are not always separate and distinct. Often one affects the other. There are times, for example, when the specific differential effect of a stressor modifies the non-specific effect. As a case in point, an injection of insulin lowers the blood-sugar level. That is its specific effect. At the same time, the non-specific (stressor) effect of insulin is to increase the blood-sugar level. Though it may be difficult to detect the more minute non-specific effects in terms of blood-sugar, the stress response may still be recognized because at high levels insulin will cause thymicolymphatic involution and adrenal cortical stimulation.

Although the concept of non-specificity is fundamental to Selye's theory, there are an increasing number of stress researchers who differ and feel that the stress response is not totally non- specific. John W. Mason (1971,1975) and Richard Lazarus (1969), for example, consider that the stress response varies depending on whether the stressor has a psychological component to it or is purely physiological. They feel that Selye's triad is mainly activated by emotional or psychological arousal and that the reason it is so common is explained by the fact that it is often difficult to separate the primary effects of purely physical agents such as heat, cold, trauma, exertion, etc.,from the psychological or emotional reactions that are attached to them-- even in laboratory animals. Mason considered this relationship, in which emotional reactions influence physiological stress responses as "psycho-endocrinology."

To Mason and Lazarus, the psychological aspects of a stimulus or an event can be the most important factor in the stress response. Burchfield (1979) goes as far as to state that stress is "anything that causes an alteration in psychological homeostatic processes." Even Cannon (Cannon & Paz 1911, Cannon 1928) acknowledged the importance of emotional factors. Though Selye touched directly and indirectly on many psychological factors, he did not sufficiently emphasize or differentiate the relative importance of purely physiological factors and these psychological influences. In any case, the psychological factors involved in the stress response will appear in many of the themes that we will be discussing.

Chapter 3

The Role of Adaptation

Adaptation is the main element of the stress response. The fact that the system is initially shaken up by an event or interaction is of less importance than how the organism attempts to come to terms with the situation. In essence, adaptation is what the stress response is all about.

Adaptation is a biological mechanism. Its purpose is to preserve equilibrium, to counteract the effect of the stressor and to minimize and delimit these effects to the smallest area capable of meeting the requirements of the situation.

The General and Local Adaptation Syndromes

Selye's major contribution to stress theory is the three-stage General Adaptation Syndrome (GAS). It is called general "because it is produced by agents which have a general effect upon large portions of the body." It is called adaptive "because it stimulates defense." It is a syndrome "because its individual manifestations are coordinated and even partly dependent upon each other."

Equally as important as the GAS is the LAS or Local Adaptation Syndrome. Local, in this instance, represents the action of specific stressors which have only a localized effect on specific parts of the body. An insect bite, for example, is a local stressor as is a toothache, a cut, or the reading of a book.

The general and local adaptation syndromes are closely interdependent. General stress can influence local stress reactions. A local stress, if strong enough, can trigger or aggravate a general stress reaction. And, by being additive, many separate and distinct local stress responses can cause or aggravate a general stress reaction.

These adaptation syndromes (the GAS and the LAS) are composed of three stages: (A) the Alarm Reaction, (B) the Stage of Resistance and (C) the Stage of Exhaustion (see Figure 2).

The Alarm Reaction

The Alarm Reaction is the body's initial response to the stressor, the "call to arms" of the body's forces. It is characterized by sympathetic nervous system activation and parasympathetic inhibition, adrenocortical enlargement and the discharge of adrenalin, an increased concentration of the blood elements, a loss of body weight and a lowering of resistance. During this stage, the heart

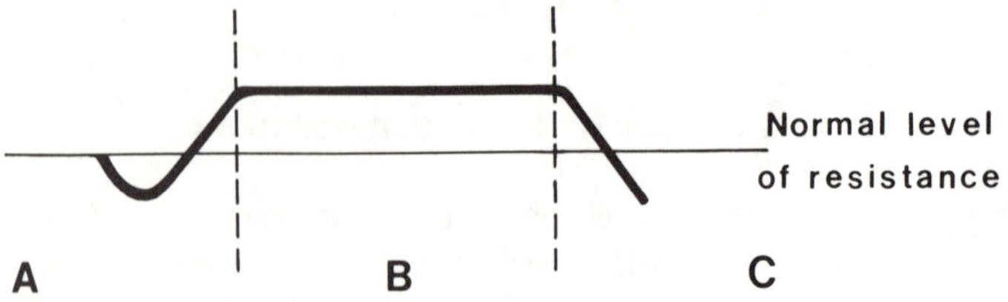

Fig. 2. The components of the adaptation syndrome

rate increases, breathing becomes shallower but faster, and the hands and feet become cooler as blood flows from the skin and viscera to the brain and the skeletal musculature--all geared to respond to the alarm.

The Alarm Reaction is subdivided into two phases: the initial shock phase in which resistance lowers and the countershock phase during which resistance begins to increase toward its maximum as defense mechanisms become active.

This subdivision demonstrates that under normal circumstances the usual initial reaction to a stressor, which involves sympathetic and adrenocortical activation, occurs for a very limited time before the forces that tend to restore equilibrium come to the fore.

Incidentally, because resistance lowers during the initial shock phase of the alarm reaction,if the stressor is strong enough, it is possible for death to result.

The Stage of Resistance

If the organism survives the Alarm Reaction, it then enters into the second stage, the Stage of Resistance. As Selye ([1956] 1976) points out, "No living organism can be maintained continuously in a state of alarm." The Stage of Resistance is, in effect, the stage of adaptation to the stressor. It is the level where resistance reaches and remains at its maximum as the body takes action to localize the problem, minimize its effect and restore more normal functioning.

Where the initial phase of the Alarm Reaction is characterized by sympathetic activation and parasympathetic inhibition, the Stage of Resistance is characterized by parasympathetic activation and sympathetic inhibition. At this stage, the adrenal cortex accumulates rather than discharges adrenalin, the blood dilutes, there is a return of body weight to normal and resistance may often rise above normal.

According to Selye, what actually happens at this stage depends on the interaction of three factors: the direct effect of the stressor itself on the body, the internal responses that act to stimulate defense, and the internal responses that act to inhibit unnecessary or excessive defense.

The Stage of Exhaustion

The third stage of the syndrome is the Stage of Exhaustion. Selye hypothesized that adaptation requires energy--"adaptation energy"--and this energy, if used without respite, leads to exhaustion. In most instances, however, a simple rest or a change of scenery can restore one's resistance and adaptability.

The fact that "adaptation energy" can be restored led Selye to the conclusion that there are two operational aspects to this energy. The first is superficial adaptation energy that is always available on demand, like money in the bank which is accessible by writing a check. It is this aspect that can be restored by a rest or a vacation. The second, deep adaptation energy, is stored away as a reserve, just as one stores away stocks and bonds which can be drawn upon, but only with some difficulty.

Selye also hypothesized that adaptation energy is finite. There is only so much and no more. Depending on how much is used up and how fast, eventually the time comes when one's store of deep adaptation energy is depleted. When this happens, the adaptive mechanisms collapse leading to signs of senility and finally death. In other words, persistent exposure to stressors over a long-term basis and/or an ineffective defense system can cause the depletion of adaptation energy and the collapse of the adaptive mechanisms. If and when this happens, signs of the Alarm Reaction reappear, such as a lowering of resistance, except at this point, as the stores of deep adaptation energy are depleted, the condition becomes irreversible (Selye 1974).

Homeostasis, Fight or Flight, and the GAS

Earlier, Cannon's term "homeostasis" was defined as representing the action of the physiological processes in maintaining an internal constancy, a steady state in the face of changing conditions.

We also referred to another phrase that Cannon introduced called the "fight or flight" response and noted that it pertained to the body's ability to mobilize its defenses immediately for some form of action when faced with a decided threat, such as pain, rage or fear, whether it was real or imagined.

Over the years, homeostasis and "fight or flight" have become somewhat blurred. Some use homeostasis as an umbrella term to include the "fight or flight" response while others use the two terms almost synonymously. Although their definitions have broadened, they are not synonymous nor is one an umbrella term for the other. There is a decided difference between the two.

The "fight or flight" response, to Cannon, referred originally to a reaction to an emergency situation or key event, specifically a perceived threat. Cannon realized that what made a "threat" out of a given interaction was one's perception of the event. Today "fight or flight" is considered as a preparation for action in the face of any anticipated challenge such as a physical attack, an exam, a tennis match, a job interview, a visit to the dentist, a social event such as a wedding, or an argument. It is the creation of a specific imbalance or asymmetry of function which sets the stage for decision making on the part of the organism as to how to respond to a situation. It is the mobilization of the body's forces for the swift action that might be needed to cope with an apparent threat (Gray 1971). The "fight or flight" response is not the decision to act but an appraisal, a preparation for the decision as to how to respond. It represents a "moment of truth" as the stage is set for alternatives to be decided upon.

Homeostasis, on the other hand,is the process of restoring equilibrium. It is the organism's attempt to maintain order, balance and stability in the face of change. It represents the decision taken after the appraisal or preparation for decision making is over. It is not the preparation to cope. It is the actual coping, the adjustment, the "coming to terms" with life's interactions.

Selye's General (and Local) Adaptation Syndrome is an umbrella term that encompasses both homeostasis and the "fight or flight" response. Cannon's emergency "fight or flight" response is equivalent only to the first phase, the initial shock phase, of the first stage of the syndrome, the Alarm Reaction. Cannon's term homeostasis is equivalent to both the second phase, the countershock phase, of the Alarm Reaction and the second stage of the syndrome, the Stage of Resistance. Homeostasis, therefore, strictly pertains to Selye's adaptive response.

Paths of Adaptation

There are a number of paths that can be taken in the course of adapting or coping. The paths of adaptation are the outcomes taken in response to Cannon's "fight or flight" response or the first (shock) phase of Selye's Alarm Reaction.

Although the use of the term "fight or flight" seems to imply that the paths that coping offers are either "fight" or "flight," most researchers agree that there are many more options available. In fact, Cannon used the terms fight and flight simply to represent the endpoints or extremes of a continuum. The following four categories of responses are offered as a reasonable synthesis of current ideas on the subject:

1. catatoxic (or fight)
2. syntoxic (or flow)
3. flight
4. fright

As a response category, catatoxic (or fight) represents active defense. It entails fighting the so-called enemy. It is represented by actual combat, aggressiveness, competitiveness, increased striving, working harder, being more argumentative or simply complaining more. Catatoxic body chemicals, for example, compel the body to defend itself vigorously with the goal of destroying the aggressor.

Syntoxic (or flow) comes from the Greek for "together" and represents putting the situation in "neutral." It entails appropriately responding to the enemy, tolerating him, living with him in a kind of symbiotic relationship rather than attacking him. Syntoxic body chemicals, for example, the anti- inflammatory corticoids, instruct the body not to fight. The syntoxic approach then is one of peaceful coexistence. Incidentally, the terms catatoxic and syntoxic were first used in relation to the stress response by Hans Selye (1974,1976,1978).

Using a prolonged, concentrated nearvision task as an illustration, the catatoxic approach would be to mobilize one's resources, tighten up and strive to do what has to be done regardless of the possible effects it might have on comfort, eye structure or efficiency. A catatoxic (or fight) approach is a goal-directed one, usually achieved with a vengeance.

A syntoxic approach to the same task would be to prepare appropriately for the work, get it done, but in due time, without any tension. It could involve taking breaks, skimming when needed and even include periodic daydreaming, all without pressure or guilt.

The difference then is one of attitude. The catatoxic response "attacks" while the syntoxic response accepts and works appropriately within the system.

Flight represents avoidance. It is an active escape from the so-called enemy. It is represented by actively withdrawing from relationships and includes avoiding working, thinking and exercise. Flight is also often manifested by a tendency to sleep or stay in bed more, even when one is not tired.

Many people seem to rationalize when they take the avoidance route. In terms of nearpoint work such as studying, one might say, "I don't need to study. I'm doing fine the way I am." Another might say, "I really have no time to study. I'm involved with so many other activities such as sports." A third might say simply, "I just don't like studying and that's that."

Fright represents a complete paralysis of action. It is where one "freezes" and does not respond at all. It is a radical form of inaction, of immobilization. In effect, fright is a severely inwardized form of flight that is usually associated with hopelessness and the feeling that there is no escape. It is the complete absence of any drive to cope with a situation. It is apathy at its worst.

There are positive and negative aspects to all these paths of adaptation. For example, being aggressive may involve gaining some control over a stressor as happens when one fights back after being insulted. However, fighting back may also result in one getting physically hurt if the other party decides to be aggressive in turn. In addition, becoming aggressive has also been known to precipitate heart attacks.

Avoiding a situation may eliminate the need to adapt further but it can also perpetuate the initial problem. If someone is afraid of flying, keeping off airplanes does not eliminate the fear.

A syntoxic approach may not always be the most appropriate response either. If one puts the treatment of a physical problem on "hold" by passively deciding to live with it, one may resist seeking professional help when such help could be beneficial. This same attitude may also tend to keep one calmer and more psychologically in control, which often allows one's natural healing processes to operate better. However, pretending that a situation does not exist purely out of fear may only tend to prolong or aggravate the condition. In fact, there are many who think that fear often tends to attract exactly what it is that is feared.

Chapter 4

Influences on Adaptation

Although stress is associated with all activities and is fundamental to life, it is obvious that a given stressor does not always affect different people the same way. Nor does it produce the same response in the same individual at all times.

The way people react and adapt is related to a number of interacting factors (Elliott & Eisdorfer 1982). First are the variables concerned with the different kinds of stressors (the type of stressor, its persistence, its suddenness and its intensity). Second are the reactions to the stressor and whether or not they become secondary sources of stress. Third are the psychological effects of the consequences and how they are evaluated by the individual as well as by those around him. Fourth are the numerous factors that mediate and modify the stress response such as genetic determinants, concurrent individual and environmental conditions, one's psychological expectations or preset, and the effects of prior conditioning. All these elements and their relative importance will be considered in this and succeeding chapters. They comprise the major influences on the process of adaptation.

Stressor Variables

It did not take long for Selye to realize that although the stress response is associated with all activities, the type of stressor that is involved is highly important in terms of its end result. Specifically, Selye recognized that there are certain stressors that are apparently pleasureful, positive, constructive, enhancing and uplifting. Some are absolutely essential to life. They literally tune one for action. Selye called the effect of these stressors "positive stress" or eustress. The prefix eu comes from the Greek meaning "good," as in euphoria.

Other stressors tend to be unpleasant, negative, destructive and damaging in nature. They place more of a demand on the body. The end result of these stressors Selye called "negative stress" or distress. Since coloquially the term "stress" has always had a negative connotation, what most people refer to as stress, Selye would call distress--stress that is negative, unpleasant and destructive.

Besides the type of stressor and its relation to eustress and distress, Selye found that persistence is an important variable. A persistent stressor can result in a prolonged and unvaried stress response which, in turn, can precipitate distress.

In addition, a stressor that is sudden and unanticipated tends to cause more of a stress response than one that is expected.

Selye also found that whether a stressor is pleasant or unpleasant is of relatively less importance than its intensity. Extreme joy can cause as much of a stress response as extreme sorrow and more of a stress response than mild sorrow. Selye liked to use the example of a mother who was informed that her son was killed in battle. Years later the son walked into the house unexpectedly. In the first instance, the effect on the mother was extreme sorrow. In the second, it was extreme joy. Yet, although the stressor types were completely different, the stressor effects (the non-specific aspects of the response) were similar because of the intensity of both situations.

Therefore, in terms of intensity, the less intense the stressor, the less of a stress response is elicited. The least stressful condition of all to Selye is indifference (see Figure 3).

Concurrent Conditions

There are many factors that affect the stress response, some of which are only indirectly related to the stressor. One of these is

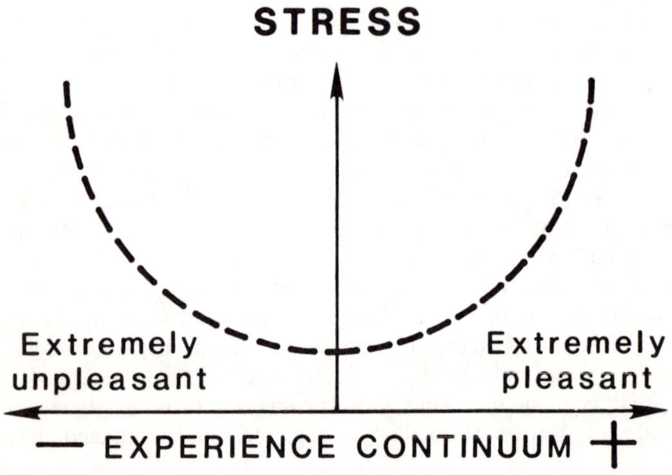

Fig. 3. The experience continuum

the current status of the individual. This includes the genetic or inherited determinants which may tend to make one more susceptible to the effects of a given stressor. This also includes the effect of developmental deprivations, one's past physical and mental history (the effect of past disease, trauma and nutritional deficiency) and one's current physical and mental state (the effect of current illness and diet).

Another series of factors that mediates the stress response concerns external environmental conditions. This includes climate (temperature, humidity, precipitation, etc.), air pollution (smog, smoke, fumes and irritants, etc.), water pollution (phosphates, dioxin, etc.), food pollution (additives, colorings, etc.) and concurrent ongoing events (crowding, noise, emotional outbursts, etc.).

Selye also highlights the importance of the type of adaptive demand, whether it is "developmental" or "redevelopmental." A developmental adaptive demand represents a situation where a system that is already adapted is required to increase its activity. An example of this is forcing a muscle system to perform more than the usual work required of it.

A redevelopmental adaptive demand represents a situation where a system, organized for one type of function, is forced to readjust itself to an entirely different kind of activity. An example of this would be a binocular visual system geared to operate in a three-dimensional space world being compelled to view flat, two- dimensional surfaces such as a television screen, a video display terminal or the printed page for prolonged periods of time.

Psychological Factors

Selye also recognized that one's personality type plays a decisive role in the eustress/distress continuum. Though there are many aspects to what is called personality, Selye loosely classifies human beings into two types, either "racehorses" or "turtles." Racehorses are those who thrive on action and pressure and are happy only with a vigorous, fast-paced lifestyle. Turtles are those who require peace, quiet and a generally tranquil

environment in order to be happy. To a turtle being in a quiet, subdued situation is eustress. To a racehorse it is distress. To a racehorse being in an active, hustle and bustle environment is eustress. To a turtle it is distress at its worst.

Selye also realized the importance of attitude. He saw distress as coming from two main sources: (l) a stress response that is prolonged and unvaried due to a persistent stressor, and (2) frustration. Frustration can be considered the result of reaching a psychological impasse, where a desired goal appears unreachable. However, what is an impasse for one person is often not an impasse for another. It is apparent then that frustration is more the result of one's attitude to an event rather than the event itself. It also becomes obvious that one's attitude can either convert a positive stress (eustress) into distress or a potential distress into eustress. Understanding attitudes, therefore, is fundamental to understanding the various stress effects that result from given events.

In addition to personality and attitude, there are many other psychological factors that influence the adaptive response. These include:

value system: the goals and priorities that determine one's needs and motivations
ego strength: one's self-concept or self-image
perceived controllability: the perception as to whether or not an event offers a way to cope
perceived predictability: the perception as to whether or not one can predict the outcome of a situation

Prior Conditioning

One of the more important influences on adaptation is prior conditioning. This pertains to how one has adapted to previous situations and how automatic this response has become. Conditioning is a way the past casts its influence onto the present. This occurs in a number of ways. In one instance, it is seen as a response that is automatically and repeatedly elicited although alternatives are potentially available. It is also noted in the repeated and automatic skipping or bypassing of certain steps in an expected sequence.

The stress response, just like any other response, can become conditioned based on previous reactions to stressor agents (Morse & Furst 1979, Nurenberger 1981). In terms of the stress response, there are two major conditioning styles: (1) sympathatonic (or sympathicotonic) and (2) parasympathatonic (or vagatonic).

The first phase of the Alarm Reaction (the shock phase) typically involves a sympathetic/adrenergic response. In sympathatonic conditioning, this basic response to the stressor is prolonged, sustained and augmented. It is a forestalling of a more normal adaptive reaction through a prolongation of the first phase of the Alarm Reaction. Individuals who are sympathatonic tend to be more prone to cardiovascular diseases such as hypertension, rapid heart beat and heart attacks.

The normal adaptive reaction (the countershock phase of the Alarm Reaction and the Stage of Resistance) involves a parasympathetic/cholinergic response. Parasympathatonic conditioning represents the bypassing of the initial sympathetic response which, if present, is weak and of short duration. The parasympathetic reaction in these individuals is usually extreme and occurs rapidly. Often it is called "parasympathetic rebound." Parasympathatonic conditioning is the most common conditioned stress response, causing a great variety of dysfunction including low blood pressure, asthma, ulcers and arthritis (Gillilan 1954). Shute (1979) considers that an overreactive parasympathetic/cholinergic response is typically precipitated by a "weak" nervous system, one that reacts more strongly to stress.

In either case, persistent sympathatonic and parasympathatonic conditioned responses are inappropriate and maladaptive reactions. They are imperfections of homeostasis or a distortion of normal homeostasis. As such, they often either lead to or become "diseases of adaptation." An example of this is the allergic reaction to a given stressor (e.g., ragweed) which, if excessive, is experienced as the actual disease. As Selye (1976) has pointed out, "stress can either cure or aggravate a disease, depending upon whether the...response...is necessary or superfluous."

Diseases of adaptation encompass many diverse conditions such as:

- gastrointestinal disorders: gastric or duodenal ulcers, colitis, diarrhea, indigestion, etc.

- cardiovascular disorders: hypertension, arrhythmias, migraine, etc.

- respiratory disorders: allergy, asthma, hyperventilation, etc.

- musculo-skeletal disorders: arthritis, bursitis, rheumatism, etc.

- skin disorders: eczema, acne, urticaria, psoriasis, etc.

- immune system disorders: involution of the thymus and lymph nodes.

- psychological disorders: anxiety, insomnia, depression, impulsivity, decreased concentration, dizziness, tics, stuttering, nightmares and various forms of addiction (alcohol, drugs, smoking), etc.

- visual disorders: functional myopia, accommodative and convergence breakdowns, etc.

This list should make it easy to realize that though adaptation is supposedly essential for restoring equilibrium and establishing some form of operational normalcy, it often can become a problem (Wittkower & Warnes 1977, Henry & Ely 1976). It is not an exaggeration to state that most of our persistent and chronic-type diseases tend to be diseases of adaptation. To put it another way, persistent illness, disease and distorted function are often the result of an almost addictive form of unsuccessful adaptation. They are mal-adaptive responses.

If a response is adaptive, meaning that for survival purposes an organism can modify its responsiveness either in quality, quantity or direction, a response also has the potential to become maladaptive. This is the negative aspect of flexibility. An organism can overrespond to a given situation. It may be an adequate response for a given system but throw other systems off balance. A response may be adequate initially for a particular situation or event but become conditioned so that an organism continues to operate the same way even when the provoking situation returns to normal. The response can persist inappropriately so that the response itself becomes the problem as in the case of hypertension or allergy. In addition, we may learn to tune out responses so that we do not react to certain situations at all. While adaptive responses in general, according to Vander (1981), often represent a "trade-off in which one variable is sacrificed in order to regulate a more critical variable," maladaptation is a form of carrying this trade-off to an extreme so that the negative aspects begin to outweigh the benefits of the organism.

The Optimal Stress Level

Health, performance and well-being have also been related to Selye's concept of eustress and distress. Everly and Rosenfeld (1981) hypothesize that everyone has an optimal stress level (see Figure 4).

This level is "the apex of one's tolerance for stress as a productive force." As long as the degree or intensity of stressor agents and the resulting Alarm Reaction (stress arousal) remain below this optimal level, health, performance and general well- being remain positive (eustress). As the degree of intensity of stress arousal increases beyond this hypothetical optimal level for a given individual, eustress becomes distress, resulting in an increasingly negative and deleterious effect on health, performance and general well-being.

It should become fairly obvious how the stress response can be a reasonable indicator of the intensity of one's life and one's

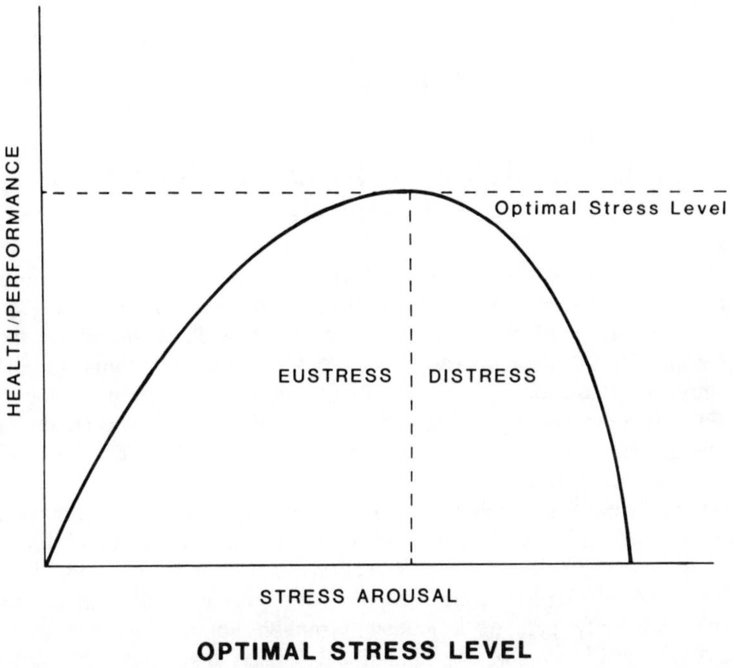

Fig. 4. The optimal stress level

ability to cope. As Selye (1976) has put it, stress "parallels one's life" representing the "wear and tear of the body caused by life at any one time." By noting all the different influences on adaptation, we can also begin to see, as Selye noted, why "the same stress which makes one person sick can be an invigorating experience for another."

Chapter 5

The Physiological Basis of The Stress Response

To do justice to an analysis of the stress response requires touching on some of the key physiological mechanisms that appear to be involved. One reason is to provide a flavor of the complexity of the process. Too often we tend to think of the stress response on a very simplistic level as either involving a sympathetic-parasympathetic nervous system action or the action of one or two key hormones. It is far more complicated than that. And current research in the area is only just beginning to scratch the surface of this whole process. A second reason is somewhat related to the first. It aims at providing some background information so that when specific neural or biochemical interactions are referred to in future chapters, the reader will realize that they are only part of a greater symphony of action.

Any classification of the operational aspects that compose a total process is bound to be arbitrary yet, through its limitations, elements of the whole may be sensed. In addition, intentionally or not, most classifications tend to establish a functional hierarchy of sorts. This survey will not be an exception. One consolation is that assessing priorities is often a valuable teaching tool. In the case of the stress response, most researchers agree that, at least in terms of current knowledge, there are some interactions that appear to be more fundamental, or at least more important, than others.

The Hypothalamus

The universally acknowledged primary element in the stress response is the hypothalamus (Levine 1971). It is the hypothalamus that appears to initiate and coordinate the entire stress/adaptation reaction. The hypothalamus receives information from many sources. It receives information from the brain stem regarding the basic autonomic life-support functions such as heart beat, respiration and cardiovascular functions. It receives information from the reticular formation regarding states of wakefulness and attention. It receives information from the other aspects of the limbic system of which it is a part, such as the amygdala, the hippocampus and the pituitary gland. It also receives information from the action of the cortex which is constantly engaged in some form of sensory decoding, motor control, sensory-motor integration and feedback, imagery, verbalization, problem solving and recall activities. In turn, the hypothalamus acts like a "kind of first sergeant, passing on orders from the brain to the rest of the body and seeing to it that they are carried out" (McQuade 1974). Anatomically, the hypothalamus is composed of two reciprocally inhibitory zones (Bovard 1961). One zone, the posterior-medial, controls sympathetic nervous system function and stimulates hormone production. The second zone, the anterior-lateral, controls parasympathetic nervous system function and the inhibition of hormone production. The importance of the hypothalamus rests in its ability to affect both autonomic nervous system function and the operation of the endocrine system, mainly by its ability to activate the pituitary gland (see Figure 5).

The génerally accepted scenario, which may not be totally accurate, is that the hypothalamus reacts to a stressor by first activating the sympathetic nervous system and then, if the stressor persists or is a massive one, activating the pituitary gland. Since the neural route appears to be the first line of action, let us start by looking at that pathway.

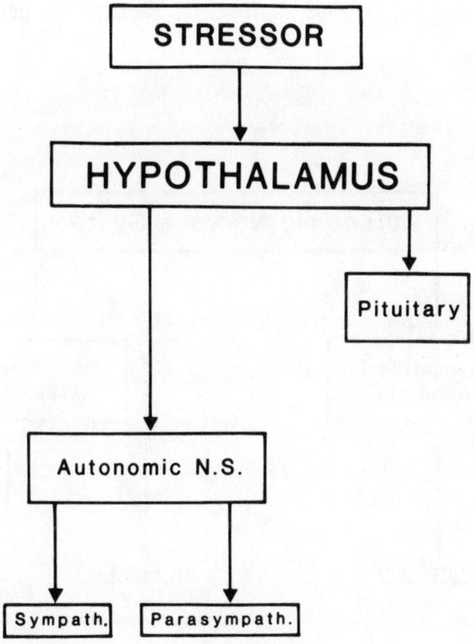

Fig. 5. The action of the hypothalamus

The Neural Route

The main neural mediator of the stress response is the autonomic nervous system which consists of two subdivisions, the sympathetic and the parasympathetic. These act as antagonists. If one is in ascendance, the other is inhibited (Gillilan 1954). (see Figure 6).

The sympathetic division, under ordinary circumstances, is the initial system activated by a stressor. It tends to be the dominant system when a massive or rapid response is required. It increases heart rate, respiration, blood pressure and sweat gland activity so that the organism can respond more appropriately to the situation at hand.

The parasympathetic division tends to be dominant when there is a need to restore or maintain normality. It provides a more stable heart beat and respiration when the body is generally quiet and at ease. It also provides a mechanism for resisting change.

Each of these pathways, the sympathetic and the parasympathetic, is composed of two major neurons (Marx 1979). The first neurons of each originate in the brain stem or the spinal cord and extend to the periphery. These first neurons (of both

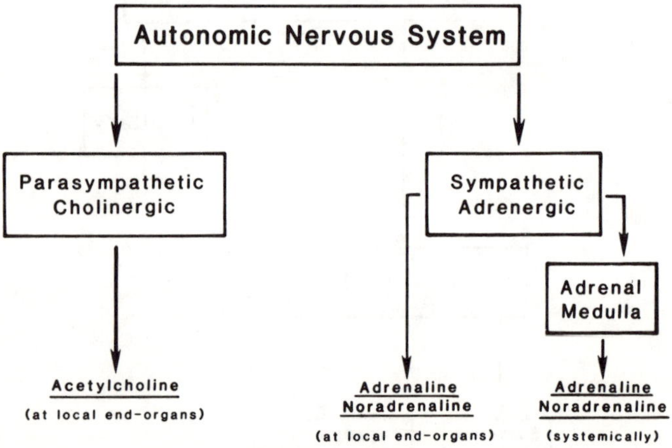

Fig. 6 The autonomic nervous system

divisions) tend to be cholinergic, using acetylcholine as a neurotransmitter. The second neurons, which form the actual contact with the specific target organs, differ. The second neurons of the parasympathetic tracts are cholinergic while those of the sympathetic tracts tend to be adrenergic, secreting catecholamines (noradrenalin or norepinephrine) as neurotransmitters (see Table 1).

Table 1.

The Sympathetic and Parasympathetic

Nervous System Tracts

	Neuron No. 1	Neuron No. 2
Sympathetiac	C	A
Parasympathetic	C	C

The adrenergic system can be both focal, through its secretion of catecholamines at the local end organs, and divergent, through its effect on the adrenal medulla, which releases catecholamines systemically. This dual action enables the sympathetic/adrenergic aspect of the stress response to reach many target organs almost simultaneously. It is this dual action that gives the alarm stage of the stress response the neuroendocrine characteristics that Cannon felt were fundamental to the "fight or flight" reaction (see Figure 6).

The cholinergic system, unlike the adrenergic system, is highly convergent with its main effects localized at the individual organs. This is one of the reasons why not all systems are restored to normality at the same time after a stressor is removed (see Figure 6).

The Hormonal Route

The hormonal route deals with hormones. Hormones, according to Selye, are "chemical messengers which carry instructions in a code which only certain organs can read" (Selye 1976).

The hormonal aspect of the stress response, similar to the neural aspect, starts with the activation of the hypothalamus by the stressor agent. The hypothalamus then responds by secreting the following key series of hormones which act as "releasing factors" activating the pituitary gland in very specific ways (Ciaranello 1983, Shute 1979) (see Figure 7):

1. CRF (Corticotropin Releasing Factor):stimulates the anterior pituitary to secrete ACTH
2. TRF (Thyrotropin Releasing Factor): stimulates the anterior pituitary to release thyrotropin
3. SRF (Somatotropin Releasing Factor): stimulates the anterior pituitary to release somatotropin
4. ADH (Anti-Diuretic Hormone, vasopressin): stored in the posterior pituitary and released in response to stress.

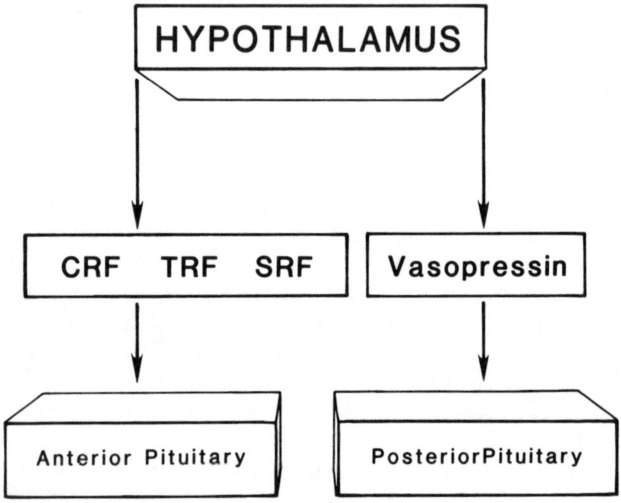

Fig. 7. The hormonal action of the hypothalamus

The pituitary gland (see Figure 8) is anatomically an extension of the hypothalamus. It is divided into two lobes, an anterior lobe which is composed mainly of glandular tissue and a posterior lobe which is composed mainly of nervous tissue.

The major stress hormone of the posterior pituitary is vasopressin which it releases into the blood stream (see Figure 9). Vasopressin works in two ways. It contracts the arterial walls which results in an increase in blood pressure and it maintains the water balance of the body by increasing the permeability of the collecting ducts of the kidney which results in water retention.

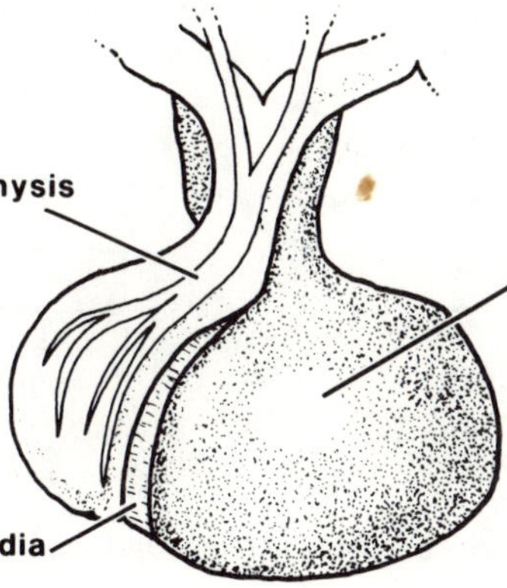

Fig. 8. The pituitary gland

The anterior pituitary produces the following six major hormones (see Figure 9 and Figure 10):

1. Three types of gonadotropin (GTH): regulate reproduction
   a. Follicle-stimulating hormone (FSH)
   b. Interstitial cell-stimulating hormone (ICSH)
   c. Luteinizing hormone (LH)
2. Somatotropin (STH): regulates growth and increases the concentration of fatty acids and glucose in the blood
3. Thyrotropin (TTH): regulates the thyroid
4. Adrenocorticotropin (ACTH): regulates the adrenal glands

The last two, thyrotropin and ACTH are the anterior pituitary hormones most concerned with the stress response.

Thyrotropin (TTH) acts on the thyroid, stimulating the production of thyroxine which increases metabolism. ACTH acts specifically on its target organ, the adrenal cortex, stimulating the synthesis and secretion of over thirty-six different corticoids. These

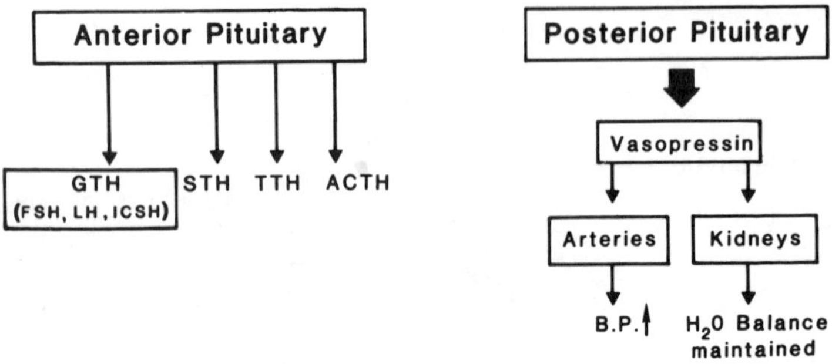

Fig. 9. The secretions of the anterior and posterior pituitary

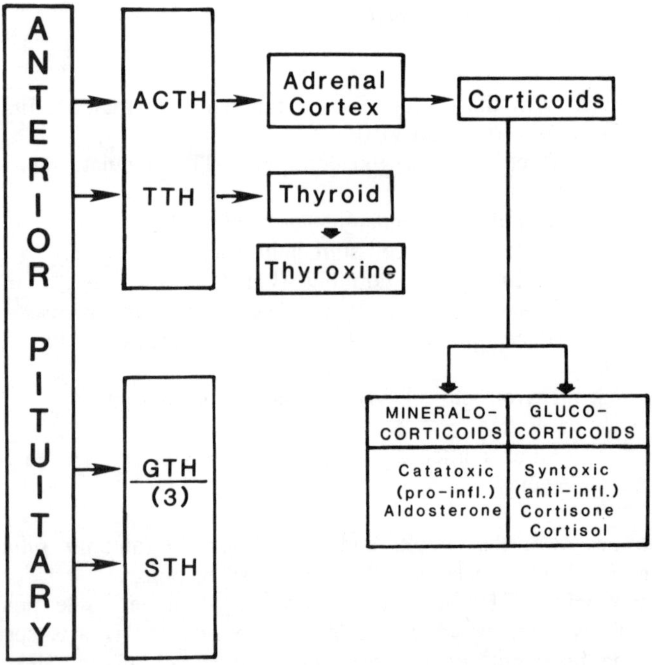

Fig. 10. The action of the anterior pituitary

corticoids are usually subdivided into two groups: mineralocorticoids and glucocorticoids.

Mineralocorticoids, secreted by the center layer of the adrenal cortex, mainly influence mineral metabolism, primarily the retention of sodium and the excretion of potassium. Aldosterone is a major example of a mineralocorticoid. The mineralocorticoids are pro-inflammatory of catatoxic. They isolate and fight the stressor. They also operate best in mild or local stress situations and are of less benefit in cases of systemic or general stress situations.

Glucocorticoids, secreted by the middle layer of the adrenal cortex, increase blood sugar, which is how they get their name. These hormones, mainly hydrocortisone and cortisol, are anti- inflammatory or syntoxic. They dominate in cases of more severe, prolonged and generalized systemic problems.

Glucocorticoids act on many organs such as the kidneys, inducing renin formation which raises blood pressure (vasopressin increases blood pressure by acting on the arteries); the liver, stimulating glucose production by converting stored glycogen (animal starch) into glucose; the lungs, dilating the bronchi, increasing the gaseous exchange and causing an increase in breathing action; the pancreas, secreting glucagon (the opposing hormone to insulin) which increases blood sugar; the heart, increasing the heart rate; the peripheral vasculature, contracting the blood vessels, forcing more blood through the liver, kidneys, heart and brain, thereby reducing the loss of body heat; the adrenal medulla, releasing epinephrine (adrenalin) and norepinephrine (noradrenaline) systemically.

The remaining pituitary hormones, the three types of gonadotropin and somatotropin, decrease during the stress response. Under normal circumstances, somatotropin stimulates growth while ACTH and cortisol are growth inhibitors. Under continued stress, somatotropin production decreases while ACTH and cortisol activity increases (Gray 1971). This may account in some way for the observation that persistent stress tends to stunt bodily growth in children.

Feedback Effects

The autonomic nervous system, in most instances, has a reciprocal mechanism so that after an initial sympathetic reaction to the stressor, under normal conditions, the parasympathetic division comes to the fore. The hormonal system modulates itself based on feedback effects.

As the glucocorticoid level in the blood increases, it triggers a decrease in the production of Corticotropin Releasing Factor (CRF) by the hypothalamus which results in a decrease in ACTH secretion by the pituitary (Levine 1971) (see Figure 11).

As the thyroxine level in the blood increases, it triggers a decrease in the production of Thyrotropin Releasing Factor (TRF) by the hypothalamus which causes a decrease in the secretion of thyrotropin by the pituitary. The increase in the blood level of thyroxine also acts directly on the pituitary to decrease thyrotropin secretion (see Figure 12).

As the blood glucose level increases, it triggers a decrease in the production of Somatotropin Releasing Factor (SRF) by the hypothalamus which decreases the release of somatotropin by the pituitary (see Figure 13).

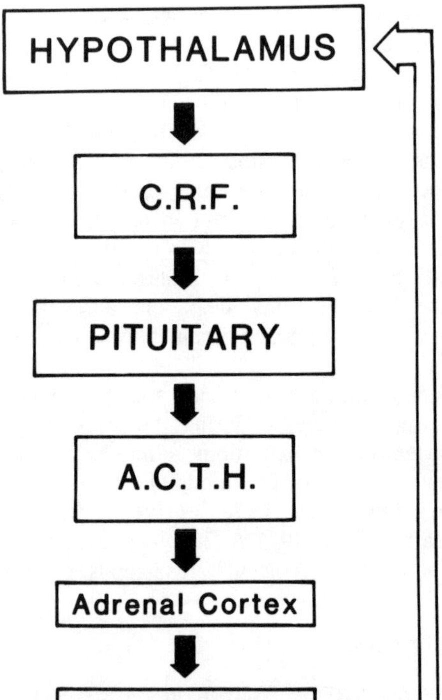

Fig. 11. Feedback effect--A.C.T.H.

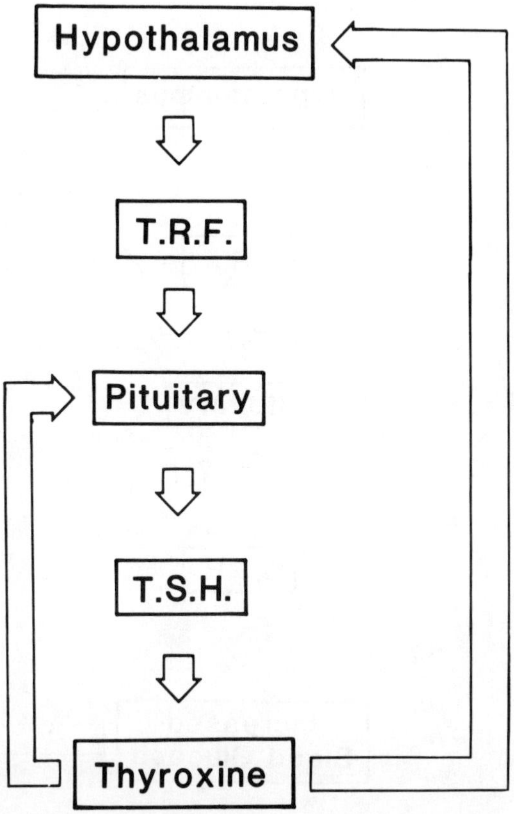

Fig. 12. Feedback effect--thyrotropin

The Cellular Route

Regardless of which route is taken, the neural, the hormonal or a combination of the two, the end result always casts an effect on the individual cell (see Figure 14).

The stressor stimulates the hypothalamus which activates the endocrine system to secrete hormones and the nervous system to secrete neurotransmitters. Both hormones and neurotransmitters are biochemicals. Both are the "first messengers" to the cell. The "first messengers" stimulate individual cells into action by activating receptors on the cell wall which then trigger "on" enzymes (adenylcyclase) and "off" enzymes (phosphodiesterase) in the cell to either increase or decrease the production of cyclic nucleotides.

The cyclic nucleotides, specifically cyclic adenosine monophosphate (cyclic AMP), are the "second messengers" in this internal relay race (Nathanson & Greengard 1977). Their action on

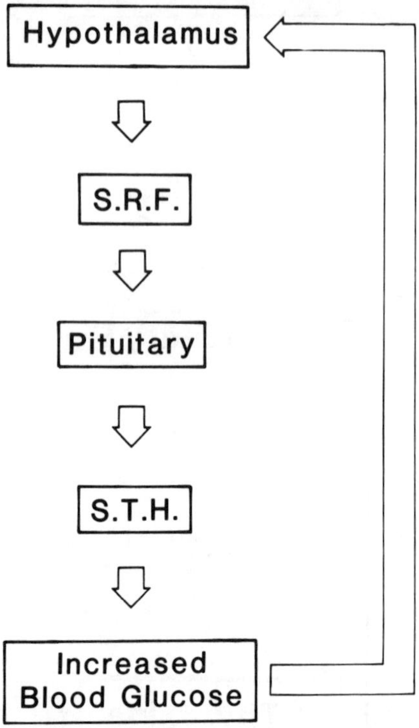

Fig. 13. Feedback effect--somatotropin

a cellular level affects general body functions, metabolic reactions, growth, air passage size, blood vessel size, heart rate, kidney function and cell division. In the initial aspect of the stress response, the "first messenger" epinephrine or adrenalin, stimulates the "on" enzyme adenylcyclase to increase the production of cyclic nucleotides. During adaptation, acetylcholine signals the "off" enzyme phosphodiesterase to break down the cyclic nucleotides.

The action of the cyclic nucleotides has been found to be interrelated with the metabolism and function of the calcium ion which acts as a parallel second messenger system. This system can modulate not only the action of cyclic AMP, it can also modulate the metabolism of fatty acid prostaglandin which then influences adenylate cyclase metabolism (Cheung 1982).

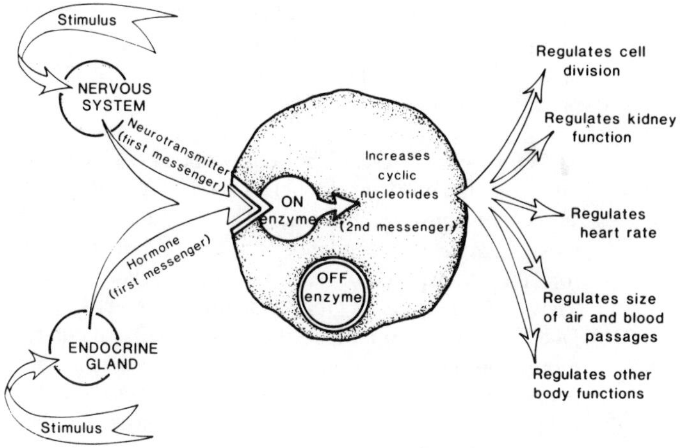

Fig. 14 Cellular Action

Stress Response Axes

In studying the stress response, certain factors become obvious. First, for any given stressor, it is rare for the neural and endocrine routes to be activated in their entirety. Second, different aspects of the stress response seem to come to the fore in relation to particular types of stressors. And third, there appears to be a temporal sequencing, a time element in the organization of the stress response. Certain physiological responses seem to come before others. These observations have led to the division of the neural and endocrine routes into the following somewhat arbitrary yet useful "stress-response axes":

1. The Neural Axis--The neural axis is represented by the action of the autonomic nervous system. Its purpose is to provide speedy and direct innervation to specific endorgans. It is the most immediate mechanism by which the organism responds to stressful stimuli. It is not fully effective, however, in mediating prolonged, chronic or high levels of stressor activity.

2. The Neuroendocrine Axis--The neuroendocrine axis aids in maintaining high levels of sympathetic-adrenergic stress arousal for more prolonged periods. It is slower acting than the neural axis because of its need to

use systemic circulation to transport the necessary hormones. This axis is primarily mediated by the secretion of epinephrine (adrenalin) and norepinephrine (noradrenaline) by the adrenal medulla. It is the axis upon which Cannon's "fight or flight" response is primarily based.

3. The Endocrine Axes--The endocrine axes, comprising three subdivisions, are usually the last pathways in the stress- response continuum. This is due to their almost complete dependence on the circulatory system for the transportation of the appropriate hormones. They also require a greater stimulus to activate them. Due to this, the endocrine axes usually mediate the more chronic and prolonged stressor actions. The endocrine axes are what are primarily involved in Selye's General Adaptation Syndrome. They consist of the following subdivisions:
   a. The Adrenal-Cortical Axis--This axis involves the release and action of ACTH and the glucocorticoids.
   b. The Somatotropic Axis--This axis involves the release and action of somatotropin (STH) into the systemic circulation.
   c. The Thyroid Axis--This axis involves the release and action of thyroxine into the blood stream.
4. The Posterior Pituitary Axis--The posterior pituitary axis involves the release and action of vasopressin.

Current Research Trends

This survey of the neuro-physiological-biochemical actions involved in the stress response has just skimmed the surface of what is really going on. New knowledge is being gained almost daily in regard to neurotransmitters, peptides, endorphins, the immune system and their relationship to the stress response.

For example, a neuroregulator is a term used to encompass any substance involved in interneuronal communication. It includes both neurotransmitters which act within a neuronal synapse and neuromodulators, such as hormones, which alter neuronal activity from greater distances from the site of release and which act through mechanisms other than synaptic transmission.

New research in this area has indicated that some neurons contain more than one substance that may be a neuroregulator and that two or more neurotransmitters may function at a single synapse (Elliott & Eisdorfer, Schmeck 1984, Miller 1982). There is also evidence that behavioral events can alter the function of neuroregulators and that altered neuroregulator function can alter behavior (Barchas et al. 1978). In terms of the immune system, it has been found that adrenal corticosteroids usually suppress the immune response to infection and other foreign tissue but occasionally can enhance it (Fauci 1975; Keller et al. 1981; Shavit et al. 1984). Acetylcholine can activate "killer T cells" which search out and destroy malignant cells while epinephrine, dopamine and prostoglandin tend to suppress immune cells that produce antibodies, leaving the body susceptible to illness (Ader 1982).

Hormones are now considered to be an evolutionary form of cell- to-cell communication (Kolata 1982). In fact, classical insulin receptors are found all over the brain as well as in the testes and the liver. Hypothalamic hormones have been found not only in the hypothalamus, the pituitary and other brain areas but in the peripheral nervous system as well. They are all peptides, a short chain of amino acids. In some instances, they act like neurotransmitters while in other instances they modulate the actions of neurotransmitters. Somatostatin, for example, can lower blood sugar levels in non-diabetics and inhibit insulin production at the same time.

Since many neurotransmitters are hormones, there is the implication that nerve cells may also be glandular in nature (Bloom 1981). Those neurotransmitters that are basically peptides have been given the name neuropeptides to indicate that they are both hormones and neural transmitters. Originally acetylcholine was thought of as being the only neurotransmitter. Ten years ago, fewer than six neurotransmitters had been discovered. Today, there are over thirty peptides that have been recognized as being neurotransmitters or neuromodulators (Krieger 1983). These include hypothalamic releasing hormones, pituitary hormones, gastrointestinal peptides and many others (Gurin 1979). One of the more important of these neuropeptides, especially in relation to the stress response, is beta-endorphin.

Peptides that mimic the action of opiates such as morphine are called endorphins or "endogenous morphine," signifying that they are morphine-like chemicals produced by the body itself. Three distinct families of opiate-like brain chemicals have been discovered: enkephalins, dynorphins and endorphins. The term enkephalin comes from the Greek "in the head." Cells containing enkephalins are widespread in the brain including the spinal cord, hippocampus, brain stem and corpus striatum. Endorphins are found in nerve cells at the base of the hypothalamus and in endocrine cells in the anterior lobe of the pituitary. Both enkaphalins and endorphins control blood pressure, body temperature, hormone secretion and body movement governance (Bloom 1981).

It has been determined that endorphins are produced in response to both physical and psychological stressors (Pert 1981). In the case of physical stress, such as trauma, the release of endorphins produces a natural analgesia effect that dulls the individual's response to pain. In terms of psychological stress, the release of endorphins produces a feeling of euphoria.

The endorphins also play a major role in the process of adaptation. Their effect is opposite to that of the initial sympathetic-adrenergic stress response. They reduce heart rate, blood pressure and respiration and also calm motor activity. Endorphins also counter adrenalin- triggered overreactions. Their action appears to mimic and support the effect of parasympathetic-cholinergic activation.

Corticotropin Releasing Factor (CRF), which triggers the pituitary to secrete ACTH, also triggers the release of beta- endorphin as part of the overall stress response (Vale et al. 1981).

Recent studies have also shown that ACTH is not only released by the action of hypothalamic CRF but by vasopressin, catecholamines and other hormones as they act on the anterior pituitary. Glucocorticoids may inhibit ACTH release by direct action on the pituitary as well as the hypothalamus. These interactions indicate that "the release of ACTH is the final outcome of the interplay among the hypothalamus, adrenal cortex, adrenal medulla and possibly other organs," an interplay which varies with the type of stress experienced and which "may act to fine-tune the responsiveness of ACTH-secreting cells" (Axelrod & Reisine 1984).

The developments in stress-response research, therefore, can be seen to encompass and integrate many seemingly diverse areas such as biochemistry, endocrinology, neurology, physiology and immunology. This integration, added to the behavioral and psychological factors that are involved, produces an intricate orchestration that in some circles has already been given the intriguing composite name of "psychoneuroimmunology" (Ader 1982).

Chapter 6

Figure-Ground and Stress

Thus far an emphasis has been placed on the stress response itself rather than on what precipitates it. We did briefly mention some stressor variables. We saw that though the stress response results from all activities, there are some stressor factors that are more important than others. A more intense demand was seen to produce more of a stress response than a less intense demand. An unanticipated or novel stressor was noted to produce more of a stress response than one that is anticipated. A prolonged and unvarying stressor tended to produce more of a response than a short-term, variable one. We also indicated that there are stressors that produce a response that is uplifting and supportive called eustress, while others produce a negative and destructive response called distress.

The question arises as to whether there are more fundamental factors underlying these observations. For example, what exactly might the "intensity" of a stressor really represent? Selye considers persistent and unvarying stressors along with frustration as being the major sources of distress. Yet, what does the fact that a stimulus is persistent and unvarying tell us about that stimulus? What are the causes of frustration? Is there a pattern, a common framework, into which most of these elements could fit?

One way that we might view many of these elements is through the larger framework of figure-ground or central-peripheral organization. There are indications that the most important stress producers appear to be those that result from key interferences in figure-ground organization at all levels, physical, physiological, perceptual and psychological. We can go one step further by noting that all life involves figure-ground organization of some form and that figure-ground organization is fundamental to human consciousness and awareness.

We inhabit a universe where existence implies awareness or responsiveness to a myriad of stimuli. We are designed to get meaning from the environment. Sensory stimulation is the matrix for gaining information from the world around us. We hone in on or "peak" some aspect of the environment in order to gain meaning. It is this "peaking" of selective fine tuning that is the essence of figure-ground organization. The "ground" merely represents unactualized "figures."

At another level, figure-ground organization represents the ability to make choices. The mass of stimuli that impinge on the varied sense receptors of the body are potential figures. Making any kind of choice entails selecting one stimulus over others for contemplation, decision-making or action. It appears that the only real potential personal freedom we may have in life is the ability to make these kinds of choices--to extract selectively figure from ground--and thus create some form of order for ourselves.

Figure-ground organization is so basic to existence that it is structured into our very being through our anatomical physiological and perceptual programs. Our anatomy, for example, decides the "name" of the living "game." Our physical structure makes certain things obstacles to us that might not be obstacles to a different organism with a different physical makeup. Whatever is an obstacle anatomically becomes a potential figure.

Sense receptors are designed to direct awareness to selected wavelengths. They are less "windows to the world" than they are limiters of information. As William James (1948) noted, our impression and knowledge of the world would be radically different if the optic nerves and optic pathway were connected to the ears while the auditory nerves and pathway were connected to the eyes. We would then hear lightning, see thunder, and see, rather than hear, symphonic music.

Similarly, motor systems also limit what we do and how we do it. If we had flippers instead of arms, we would certainly handle the world around us differently. Even the form of verbal labeling and communication we use hinders as much of our knowledge of reality as it aids in analyzing and knowing the world around us. Therefore, sense receptors and motor systems structure what elements we choose to be figure and how we then act on them.

Physiological programs set our operational bounds through built-in neurological and biochemical systems. For example, a palmer grasp is programmed into an infant's development in order for the child to be able to take hold of larger objects. Later, a thumb- forefinger pincer grasp is programmed for fine manipulation. Hormonal systems operate on genetically programmed "time" schedules. Puberty is an example of one of its effects. The menopause is another. The nature of stereopsis, of foveal versus peripheral retinal processing, and even basic sensory processing in terms of light-dark differences, patterns, regularities, spatial frequency and edge awareness are all aspects inherent in human physiology that aid in pulling figure from ground.

At a perceptual level, there are also built-in "ground rules," some of which the Gestalt school of psychology has referred to as "laws of organization" (Kohler 1947, Kaffka 1922, Wertheimer 1921,1923,1944). These are apparently designed to organize awareness of figure-ground relationships so that one perceives with closures that are common for the genus. For example, the ground rule pertaining to the perception of area states that the smaller a closed region, the more it tends to become dominant and be seen as a figure. Note in Figure 15 that as the white area decreases, it has an increased tendency to stand out as a figure.

The law of proximity states that objects that are close together tend to be grouped together. Note that in Figure 16, as the dots in each vertical column are moved closer together (B), the overall picture tends to be perceived as a series of vertical columns. Similarly, as the dots in each horizontal row are placed closer together (C), the overall picture tends to be perceived as a series of horizontal rows of dots.

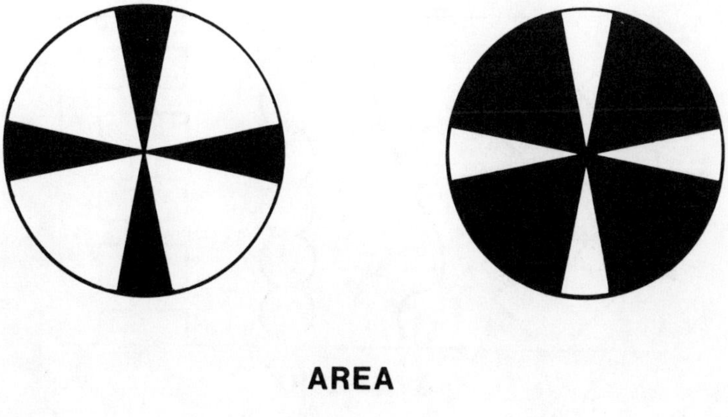

Fig. 15. Gestalt law of organization--area

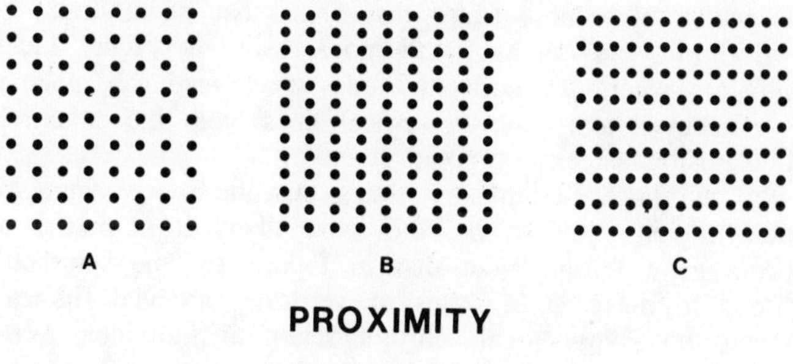

Fig. 16. Gestalt law of organization--proximity

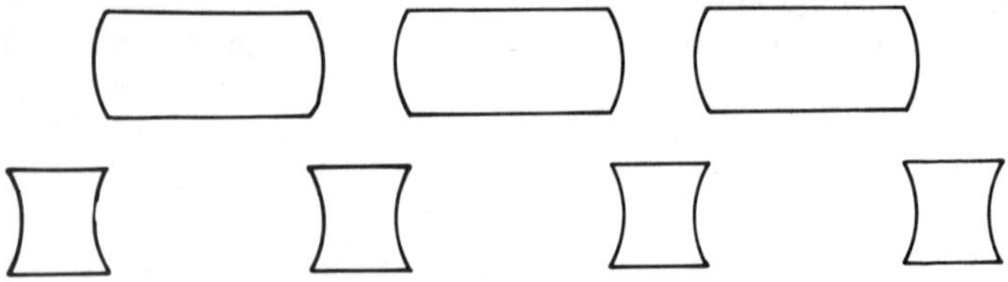

CLOSEDNESS

Fig. 17. Gestalt law of organization--closedness

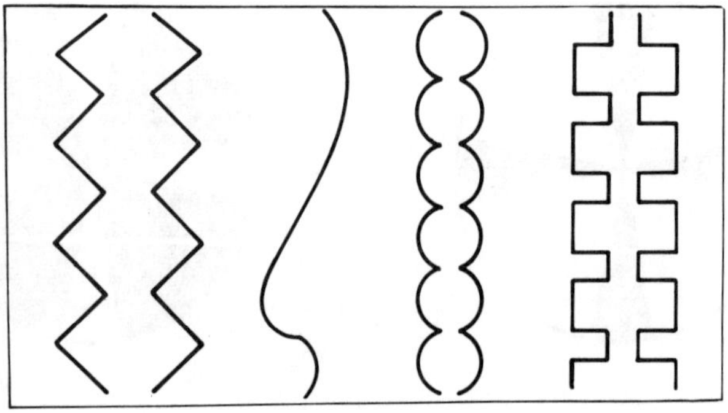

SYMMETRY

Fig. 18. Gestalt law of organization--symmetry

The law of closedness states that areas that have closed contours tend to stand out as figure more easily. In Figure 17, just manipulating where the horizontal lines are placed can make this series of convex and concave vertical lines look like either elongated television screens or apple cores.

Finally, the law of symmetry states that the more symmetrical a closed region tends to be, the more likely it is that it will stand out as a figure. Note that in Figure 18, the symmetrical lines tend to make up a series of vertically oriented figures. If they were not symmetrical, only a series of individual vertical lines of varying design would be noted.

Figure and ground are not isolated aspects. They are intimately interconnected. Figure needs ground and ground needs figure. Every figure, whether it be an object, an idea or an event must relate to a context which is its ground. Kurt Goldstein (1963) points out that a word is "understandable only within a definite context, within a definite sentence, within a certain cultural sphere".

An illustration of this is the word queen. Change the context and the figure changes. One can use the word queen in the context of female royalty. One can also conjure up more specific images if the context changes so that the reference becomes the Queen of England, the Queen of Spain, Queen Elizabeth I or Queen Nefertiti. Notice the change in the figure as the context changes to the Queen Mother, the good queen, the evil queen, the former queen, the future queen. Notice how the figure becomes even more radically altered as the figure/context shifts to Queen for a Day, Queen of the Ball, the drag queen, the queen bee and, finally, to Ellery Queen. Notice how, as the relationship between the figure and the context (ground) is altered, the meaning of the figure changes since it now has a completely different reference.

Figure-Ground and Attention

Outside of the biological mandate to organize figure in relation to ground, the most important aspects of the process involve choice, a decision on the part of the organism as to what will be figure and what relationship it will have to ground. The voluntary, selective aspect of choosing a figure for meaning from the ground can be equated with attention and attentional processes (Neisser 1976). This is what attention is all about. Attention is nothing more than the act of extracting a figure from a context, in relation to the context. It is the narrowing of consciousness as it pulls figure from ground. Perhaps an even more appropriate and encompassing definition of attention might be that it is the figure-ground methodology or strategy used by consciousness to manipulate and regulate awareness.

It is what we attend to, how we organize figure and ground and the manner in which it is done that affects our interpretation of any and all events. A group of people can see the same movie and yet have different impressions and interpretations because of the influences that affect how they attend or how they extract figure from ground in relation to the context.

Of even more importance is the fact that there is no attention without intention of some form. There is no such thing as purposeless attention. Attention is goal directed and goal oriented. It is the intention, the purpose, the end-goal that determines how attention is used.

Furthermore, the figure-ground aspects of attention demonstrate a particular area of vulnerability. In attention, for example, awareness is always divided to some degree between what can be called central, focal or discrete awareness on the one hand and peripheral, global or diffuse awareness on the other. What is usually lost or distorted in the process of attention is effective integration between the periphery (the ground) and what is attended to focally (the figure).

Attention always involves some degree of "psychic energy" or effort (Kahneman 1973) and the amount of energy used is directly related to the degree one inhibits distractions (elements of the ground) from awareness (Pribram & McGuinnes 1978). Highly focal (or central) attention in which the periphery does not play an active role or is tuned out is associated with a greater expenditure of energy than when periphery is actively involved. If attentional demand is high and prolonged, then attentional capacity or the reserves one has to maintain attention shrinks resulting in what has been called "cognitive fatigue" (Cohen 1980).

In this way, an imbalance in figure-ground (central and peripheral) processing, namely an increase in focal attention with a decrease in peripheral awareness becomes a primary producer of stress or distress because it places an increased attentional demand on the organism.

There are two other factors involving imbalances in figure-ground interaction that also appear to be major stress-producers. These are input (or information) overload and input (or information) underload.

Input Overload

Input overload pertains to the attempt to gain meaning in the presence of excessive stimulation. Excessive stimulation results from either prolonged, high-intensity stimulation or simply from too much going on at one time, either through single or multiple sensory modes. As a general rule, as the input rate increases, output will also increase. If the rate and intensity of the stimulation continue to increase, a point will be reached beyond which there will be a sudden falling off or collapse of performance and meaning as the individual becomes less and less capable of functioning adequately in terms of information processing.

Noise is a major example of input overload. Noise has been implicated as having an adverse effect on performance, behavior and both physical and emotional well-being. It can induce feelings of displeasure and annoyance, reduce both sociability and sensitivity to the needs of others and result in measurable physiological changes (Cohen 1980). Yet, in this day and age, it is almost impossible not to be confronted by some degree of noise from aircraft, cars, trucks, buses, motorcycles, stereos, radio, television, tape recorders, air conditioners, vacuum cleaners, etc. Noise acts by either overemphasizing figure in relation to ground or overemphasizing aspects of the ground in relation to what is figural at the moment. In either case, it is a disruption in efficient figure-ground interaction. There is, however, a very strong psychological component to what is called noise.

Noise specifically refers to sound that is unpleasant, annoying, upsetting or unwanted. Yet, what is noise to one person is not noise to another (Cohen 1981). Listening to loud rock music is a perfect example of this. The effects of noise can be modified by one's approach to the noise. One can suppress, tolerate or resist noise. If noise is associated with negative experiences such as failure or punishment, performance will be impaired. If it is associated with achievement, performance will tend to remain unimpaired or even be enhanced (Glass & Singer 1972). Noise then, is not totally a physical event dealing with sound. It is more often a psychological appraisal of what is being heard. In fact, noise levels alone can explain only between one-tenth and one- third of the variability in individual annoyance (Baum, Singer & Baum 1982).

Crowding is another example of input overload. There are studies that indicate that crowding appears to cause disease and early death in both animals and humans especially when there is a sudden change from being in a low-density population to being thrust into a high-density one (Dubos 1965). High-population density alone, however, does not lead to an increased stress reaction, at least not as a general rule (Freedman 1975). There are many who live in crowded neighborhoods, in crowded tenements or dormitories, who travel in crowded trains, buses, airplanes and cars and who stand in crowded lines at supermarkets and theaters. Not all of these people in all of these situations develop negative effects from the crowding.

Michael Hughes of the Virginia Polytechnic Institute and Walter Gove of Vanderbilt University, using the number of persons occupying a residence as an indicator of crowding, found that what contributed most to the negative effects of severe crowding was the type of household and the role that the person played within that household. The greatest negative effects were seen in unmarried parents, especially women, living with their children. This effect was moderated if another adult, such as a grandparent, was part of the household (Miller & Peterson 1984).

Similar to noise, crowding is a psychological phenomenon. One may or may not be in a high-population density situation to "feel" crowded. Whenever a person expresses the sensation of being crowded, he simply does not like the situation he is in (Freedman 1975). In this respect "the little house on the prairie is perhaps no better for your health than the little tenement on the avenue" (Page 1984).

One experiences the negative effect of crowding when one expects to be crowded even if a crowded situation isn't actually present (Baum, Singer & Baum 1982). Feeling that one has control over the situation also tends to reduce the negative effects of crowding (Cohen 1980; Glass & Singer 1972; Sherrod 1974, Rodin, Solomon & Metcalf 1978). It is the psychological appraisal of the situation, similar to what occurs with noise, that is the most important aspect of the crowding phenomenon. When one labels a situation as being a crowded one, the individual is actually attributing his or her dislike of the situation to others. It is a way of blaming outside factors or events for some inner dissatisfaction (Epstein 1982).

Input Underload

Input underload is often known as information underload or sensory deprivation. It represents the attempt to pull effectively figure from ground and gain meaning while external stimulation is either absent or limited. In essence, it is the withholding of information to the organism. There are three major types of input underload: sensory deprivation, social isolation and immobilization.

Sensory deprivation, in its more specific sense, refers to a limitation of stimulation to one or more of the major sensory processing channels: vision, hearing, touch, taste, smell and kinesthesia. An interesting example of sensory deprivation occurred a number of years ago. A $10 million office building was built by Lufthansa in Cologne, Germany in 1968. The designers used such ultra-efficient soundproofing that no outside or inside background noises could be heard by those working in the building. It did not take long before the employees started to complain that they could not stand the silence. The problem was finally resolved by the installation of a "noise machine" to provide a continuous background murmur of traffic, telephone and typewriter noises ("Silence is leaden" 1968).

Zuckerman (1964) found that restricting input to two sense modalities tends to be much more stressful than restriction of only one modality. Yet, sensory deprivation does not have to be maximal in order to have an effect on the organism. Prolonged attention to a single task is a form of sensory deprivation that is further magnified when it is accompanied by a reduction or suppression of non-focal input in order to maintain concentration.

More important than the physical or physiological limitation of stimulation is the psychological component. John Lilly (1972), one of the pioneers in the area of sensory deprivation, found that even in its extreme form, using the isolation tank, sensory deprivation can be minimized or maximized by pre-programming, which is why some find it upsetting while others use it for relaxation. This is another indication of the modulating effect of attitude, emotions and one's state of mind.

Social isolation refers to the limitation of stimulation that is normally provided by being with others. Zuckerman (1964), while working with the restriction of input to specific sensory modalities, found that the stress response was far greater after relatively short periods of isolation, often no longer than 1 to 1 1/2 hours. He was so impressed with these findings that he concluded that stress was primarily a function of time and isolation more so than simply a response to a novel, threatening or suggestive situation.

Studies have also indicated that field-dependent individuals demonstrate a greater stress reaction to isolation than do field-independent individuals. Field-dependent individuals kept in isolation for less than two hours showed more anxiety, body discomfort, suspicion, fantasizing and disorganized thought than did field-independent subjects (Cohen, Silverman, Bressler & Schmavonia 1961).

Social isolation does not have to be prolonged or even a real condition in order to cast its effects. Social isolation has a very strong psychological component since it is the feeling or perception of isolation that not only can aggravate the actual physical situation but can often create the effect of isolation when it does not even exist. One can be alone and not lonely. One can be in a crowd and be quite lonely. Being alone is a physical situation. Loneliness is an appraisal. Simply, each and every one of us has the ability to create our own isolation and carry it around with us. Too often we tend to close off much of the world around us and isolate ourselves psychologically. Many do it automatically when they read intently or concentrate on a particular task or job (Trotter 1973).

Immobilization refers to a restriction or a restraint of body movement. There have been studies that demonstrate that limiting general movement even for short intervals will cause changes in EEG activity (Zubeck & Wilgosh 1963) as well as an increased instability in autonomic nervous system responsiveness (Konrad &

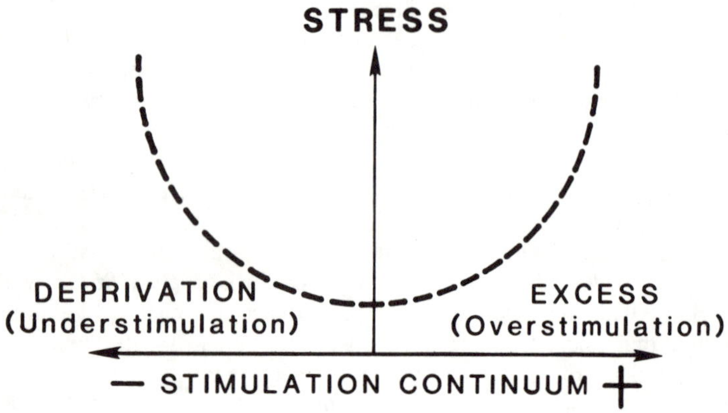

Fig. 19 The stimulation continuum

Bagshaw 1970). When visual or auditory restrictions are added to this, the effect becomes even greater.

Input overload and underload,therefore, can be thought of as representing parts of a stimulation continuum (see Figure 19). The highest stress levels are usually found at the extremes of overstimulation or understimulation. Lesser stress levels are associated with lesser degrees of overstimulation or understimulation. As Lennart Levi (1972), a major stress researcher, points out, deprivation or excess of almost any influence is stress provoking. This includes affluence as well as poverty, parental overprotection as well as parental deprivation, and extreme permissiveness as well as restriction of action.

One element that pervades most, if not all, aspects of figure-ground organization is one's perception or appraisal of the event in question. Isolation, immobilization, deprivation, crowding and noise all represent real situations, yet it is one's sense of isolation, immobilization, deprivation, crowding and noise that determines the degree of stressfulness.

Therefore, what we ourselves make focal and the degree to which we do it, what we suppress and the degree to which we do it, all in relation to the total context or ground, determines the resulting degree of stress or distress. More often than not it is we ourselves who create our own distress.

Chapter 7

The Arousal Response

Physiologists have been interested in the subject of arousal for years. Stress researchers became involved in the late 1950s when it was noticed that those who were exposed to sudden and unpredictable stressors demonstrated physiological effects that were similar to those found during the arousal response. This brought up the question as to whether or not there was a legitimate relationship between arousal and stress. To understand if a relationship does exist and, if so, to what extent, requires some knowledge about the subject of arousal.

Originally, the term arousal had a very specific connotation. Over time, however, arousal has come to represent a broad spectrum of activities. At its most basic level, arousal represents the awakening process that is necessary before one can be responsive to the environment. This level involves the transition from the sleeping state to a state of wakefulness (French 1957). At a somewhat higher level, it encompasses the way the body mobilizes for action or puts itself "on alert" to protect itself against a possible threat. At a still higher level, arousal represents one's responsiveness to a stimulus. It involves the ability to shift conscious awareness from one stimulus to another. In this respect, arousal is a fundamental part of the localization of attention.

What interested stress researchers in the area of arousal was the observation that the arousal response varies from moment to moment with the difficulty of a task as does the stress response. Arousal also reflects what an individual does, the effort he invests in doing it, his attitude toward what he is doing and his general physical and emotional state, as does the stress response. A state of high arousal reflects a greater intensity of involvement than does a low arousal state, similarly to how intensity affects the stress response.

And, as with the stress response, controllability and predictability also play a key role (Cohen 1981). Those who are experimentally exposed to controllable and/or predictable events tend to show a lower level of arousal than those who are exposed to situations that they perceive as being uncontrollable and/or unpredictable. In addition, the arousal response manifests itself by sympathetic nervous system activation as does the stress response, specifically the first phase of the Alarm Reaction of Selye's Adaptation Syndrome.

In chapter 4, it was noted that everyone has an optimal stress level that relates the stress reaction to performance and well-being. If the level or intensity of the stress response goes beyond this hypothetical level, performance and well-being tend to decrease (Everly, Jr. & Rosenfeld 1981) (see Figure 4). Similarly, many arousal theorists believe that there is an optimal arousal level that varies with the particular task on hand and which also relates to performance (Cohen 1981). Performance, for example, is seen to increase as arousal increases up to this optimal level. If arousal increases beyond this point, performance decreases. For those who equate arousal with stress, the "optimal stress level" and the "optimal arousal level" represent different perspectives of the same response.

The Physiology of Arousal

Traditionally, the Reticular Activating System (RAS) has been implicated as the major mechanism for arousal. One of its functions is to send signals to the cortex as a kind of general alarm to "arouse" the brain. It also works to maintain the individual in a conscious, aware state. In addition, the RAS controls and regulates all the motor activities of the body, voluntary and involuntary. In effect, it acts as a control center for the nervous system in general, having the ability to be quite selective in its sensitivity to different stimuli.

More current research has indicated that the arousal response is a direct function of both the RAS and the hypothalamus. This is not totally unexpected since the hypothalamus, which plays such a key role in the stress response, is considered by some to be an integral part of the RAS (French 1957).

Both the RAS and the hypothalamus directly affect autonomic nervous system function by stimulating sympathetic nervous system action. They can also inhibit parasympathetic nervous system function. This has been demonstrated by the observation that arousing stimuli can provoke pupillary dilation even when the sympathetic nerve supply to the iris is cut off (Meldman 1970). The predominant effect, however, is one of sympathetic nervous system activation. This interrelationship between the RAS and the hypothalamus may help explain why sympathetic nervous system effects can be noted in both the arousal response and in the Alarm Reaction of the stress response and why these effects are being used by many as both an index of arousal and an index of stress.

Semantics

Just as there are semantic problems with the term stress, there are also semantic problems with the term arousal. Actually, there are many similarly used words that one comes across in the literature on arousal. These include orienting (or orientation), alerting, awareness, attention and activation. It is rare to find any two authors agreeing exactly on any one of these terms. Too often they are used quite loosely and interchangeably.

Traditionally, the orienting response represented turning toward, or directing one's attention to, a novel stimulus in order to be able to deal with it (Lynn 1966). Pribram (1971) considers it to be an appropriate and automatic, non-habituated, response to an immediate action, a built-in biological element of distractability. He also perceives the orienting reflex as being composed of at least two distinct processes: "an alerting, searching and sampling component and a registration in awareness and memory of the events sampled."

Many types of stimuli can elicit the orientation response. These include a novel situation, a change in the intensity of the stimulus, a change from a drab, gray or monotonous stimulus to one that is more colorful, learned or conditioned responses to given situations, a sudden surprise, and stimuli with greater complexity or that create uncertainty, incongruity or conflict (Berlyne 1960).

Where adaptive or homeostatic reactions serve to preserve equilibrium by reducing stimulation when it is intense and increasing stimulation when it decreases in intensity, the orientation reaction always increases the sensitivity of the organism to stimulation in preparation for action (Lynn 1966).

The term alerting, to Pribram (1971), is simply a synonym for orienting. Meldman (1970) further differentiates between general alerting which serves to transform the individual from mere waking to an attentive state and specific alerting which involves either broad concentration on a single perceptual mode or a sharp focus of attention within that mode.

Awareness, at its least sophisticated level, simply implies that a stimulus or stimulus event has been registered at some level of consciousness.

The terms arousal and activation usually create the most semantic difficulty. One speaks of the arousal response and the Reticular Activating System, but not the Reticular Arousal System.

There are many who use the terms arousal and activation synonymously. Others use arousal to refer to behavioral effects and activation to refer to the physiological activity involved in the process. Humphreys and Revelle (1984) consider arousal as a conceptual dimension, a "state of the organism that in everyday terms means alertness, vigor, peppiness and activation." Meldman (1970) considers arousal as the transmission of both internal and external information of behavior. Wittrock et al. (1977) define arousal as a physiological state characterized by behavioral alertness, sympathetic nervous system predominance and desynchronized EEG activity. They define activation as the physiological process by which brain stem systems induce arousal. Beatty (1977) views activation as "a physiological concept that refers to brain mechanisms that control behavioral arousal." Eysenck (1982) defines arousal with a more psychological bias as referring to "the common effects on behavior of several apparently disparate factors such as incentives, failure feedback, ego threat, intense noise, time of day, sleep deprivation, introversion-extraversion and a number of stimulant and depressant drugs."

The Pribram Approach

Karl Pribram and Diane McGuinness (1975) offer a different and quite unique view of the arousal process. They conceive of the process as consisting of three major components. The disservice that they perform is that they add to the semantic confusion by calling one aspect arousal, another activation and the third is described as being the coordinating process between the two. In their view, arousal and activation are not synonymous terms nor do they represent physiological as opposed to behavioral aspects of the process. Regardless of the names they assign, they do offer an interesting and insightful breakdown of the elements of the process that attempts to reconcile some of the inconsistencies noted in arousal and stress research. For that reason alone, their model deserves serious consideration.

The first aspect, the one they label "arousal," is the part that registers an input or stimulus into awareness. It is the mechanism to "stop and be aware." It represents Pavlov's "What is it?" or "What's that?" reaction. Arousal is seen as essentially a phasic reaction to a stimulus. It can be said to occur when a change in input produces a physiological or behavioral change that is measurable, which implies some degree of registration of the input at some brain level. Arousal, to Pribram and McGuinness (1978), also encompasses the matching of the input with a cortical representation based on past experience since "without such matching there could be no novelty or information nor even a measure of change in intensity." Therefore, any mismatch between one's internal model of the world and a current stimulus, no matter how slight, can trigger arousal (Sokolov 1960). Pribram (1981) also indicates a strong link between arousal and the emotional or affective elements of an event (McGuinness & Pribram 1980).

Physiologically, the process of arousal is viscero- autonomic and is controlled by the amygdala and related structures of the frontal cortex. Messages from the amygdala and the frontal cortex converge on the hypothalamus and the brain stem. This, in effect, provides "forebrain" control over the "corebrain" arousal system. Actually, there are two reciprocally acting amygdala systems. One facilitates while the other inhibits arousal.

The second aspect, the one labeled "activation" is the part that represents the motoric and perceptual readiness to respond. It is the trigger to "go and do." Where arousal produces a phasic increment in neural activity, activation is tonic in nature. Where arousal represents the "What's that" reaction, activation represents the "What's to be done" reaction. Where arousal is viscero-autonomic, activation is somatomotor. Where arousal is controlled by the amygdala and frontal cortex structures, activation is under the control of the basal ganglia.

To Pribram and McGuinness (1975) "the orienting reaction involves arousal but no activation; vigilant readiness involves activation but no arousal; the defense reaction involves both arousal and activation; when neither arousal nor activation is present, behavior is automatic...."

The third aspect of the process is the one that coordinates arousal and activation. Neither arousal nor activation involve effort. Effort is involved only in the regulation and coordination of the two. It is generated not by basic awareness or the preparation for action but by the coordinating activity that goes on between them, which, neurologically, is under the control of the hippocampus. To Pribram, its major cause is inhibition. Attempting to maintain a state of non-distractability requires a great amount of effort. This means that it takes effort to maintain attention since attention requires dampening distractions. Conversely, by increasing one's willingness to be distracted, effort can be reduced and relaxation increased (Pribram & McGuinness 1975).

These three aspects of what is more generally known as the arousal process have interesting implications. One is that maintaining vigilance, or a motoric readiness to respond, in the presence or absence of external stimulation (activation) is a fundamentally different function than being perceptually oriented to an external stimulus (arousal). A second implication is that different individuals can have preferences toward being either more motor-vigilant (which is a tonically active mode) or being more novelty-oriented (perceptually responsive). Pribram also views the motor-readiness aspects of activation as directly affecting one's motivation while arousal, as noted before, is seen as primarily supporting emotion. Arousal may be one method then by which the brain is primed to respond to events in terms of how affectively interesting those events are rather than just on the basis of informational novelty. Finally, in terms of brain lateralization, Tucker and Williamson (1984) feel that the neurotransmitter substrates of activation are predominantly left hemispheric in nature while perceptual arousal is mainly a function of the right hemisphere. This leads them to speculate that right hemisphere function, associated with arousal, "may be intrinsic to a receptive mode of attention and may support holistic perception." Left hemisphere function, associated with activation, is viewed as relating to that form of internal control and vigilance that is seen in focal attention, encompassing an analytical cognitive approach to visual perception (Kinsbourne 1974, Tucker 1981).

In a broader sense, arousal, activation and the effort that results from their interaction appear to be but different aspects of the more fundamental process of attention.

Chapter 8

Attention, Arousal and Heart Rate

We have seen that both the initial stress response and the arousal/activation complex involve the action of the autonomic nervous system, mainly sympathetic stimulation. Many techniques have been used to monitor these effects. The traditional procedures involve measurement of blood pressure, heart rate, galvanic skin response and EEG. Stress researchers also use blood tests to measure various aspects of hormone production. Of all these probes, the most common technique used has involved the measure of heart rate: cardiac acceleration and cardiac deceleration.

Since arousal, activation, effort and the stress response are all integral aspects of attention, it might be expected that changes in the status of attention would cause corresponding changes in heart rate.

Attention has been considered as a "selectivity of response" (Hebb 1961) or "focused energy." (McKim 1980). It has also been viewed as a localization of conscious awareness as the brain responds to a stimulus in order to get meaning from it or do something with it. Attention has been referred to as the mental equivalent to using a searchlight to see something in the dark or the fine-tuning of a radio receiver to get better reception. It has been used as a designation for the intensity in the way one does a task (e.g., "paying attention"), for the ability to select one aspect of the ground for meaning, for alertness and for vigilance (Meldman 1970).

William James ([1890] 1948) saw attention as being composed of three aspects: a selective process by which we select what we will extract for meaning, an intensive process in which effort is used, and a sustaining process in which concentration is maintained even during prolonged monotonous tasks.

Meldman considers three components of attention: primary--an immediate reaction to a novel stimulus; secondary--a conditioned response based on past experience and tertiary--a subliminal registration of a stimulus. He also divides attention into the following types: active (a voluntary, selective directing of awareness), passive (a generalized awareness of the surround), concentration (focusing of consciousness with effort), external, internal, ideational, reflex, diffuse or global (a broad focus of attention), and focal (a limited focus of attention).

Earlier we viewed attention more encompassingly as part of the total figure-ground phenomenon, the organismic act of extracting figure from a context in relation to the context. We also considered it as a key strategy used by consciousness to manipulate and regulate awareness. As a total organismic process, we would also expect it to be reflected in neurological, biochemical and general physiological function.

Since the way arousal is usually defined (as an integral part of the attention process) and since arousal and the initial Alarm Reaction of the stress response both result in sympathetic activation, it could be expected that as the intensity of one's attention to a task increases so would the heart rate. This is based on the fact that a key physiological effect of sympathetic activation is cardiac acceleration, an increased heart rate. This, however, does not always prove to be the case. In many instances, the heart rate is noted to decelerate with attention. The question is, why?

Under conditions of rest, the heart rate is basically under the influence of the parasympathetic nervous system which tends to keep the heart rate low and more variable. When sympathetic innervation increases, the heart rate tends to rise and become less variable (Katona & Jih 1975). Studies on adults have demonstrated that the heart rate increases as subjects attempt to cope with difficult problems (Light 1981, Martin 1961).

Working with children, Kagan (1983) has found that the psychological state is an important factor. Those children who were more vigilant and anxious in regard to what they were looking at or listening to tended to have higher heart rates. Active vigilance is defined in this instance as "the degree to which the child actively attempts to assimilate unfamiliar and discrepant events."

In an earlier study, Kagan and Phillips (1964) found that cardiac deceleration, a slower heart rate, was a typical reaction that accompanied attending to visual and auditory stimuli in both children and adults. They noted especially that analytic children showed these reactions to a greater degree than non-analytic ones.

The results of these two studies may appear contradictory but the answer rests in understanding the difference between what is called "active vigilance" in the one study and to what is referred to as an "analytic" approach in the other. Kagan and Phillips (1964) note that analytic children "invest more attentional energy in the processing of external information." The key words in this quote appear to be "external information" and not "attentional energy." An analytical approach, to Kagan and Phillips, does not necessarily deal with the intensity or effort used in one's approach, per se, as much as it does with how the totality of external information is processed. A more analytic individual in this sense is considered to be more attentively observant to all the stimuli he encounters. Kagan and Rosman (1964) added another piece to the puzzle when they found that thinking tends to produce cardiac acceleration, an increased heart rate.

This takes us full circle to our earlier discussion as to how central and peripheral stimuli are processed. Many sources have indicated the importance of the two major attentional modes used in processing visual and other information: the global mode involving information gathering from the entire stimulus field and the focal mode involving a rather detailed analysis of a limited area, usually central, of the stimulus field (Kahneman & Effort 1973, Neisser 1967, Beck & Ambler 1972, Alwitt 1981, Forrest 1976b).

Lacey (1967) has pointed out that "the cardiovascular system is particularly responsive to the intention of the subject to note and detect external stimuli." He states that he has repeatedly found that "attentive observation of the external environment is productive of cardiac deceleration...." This implies that the arousal process does not simply reflect the intensity of behavior but the intended aim or goal of the behavior. In this respect, it is not incongruous that "mental work" results in cardiac acceleration, a sympathetic response, since "mental work" implies "thinking" in which the importance of external environmental cues are often minimized. That is to say, internalized problem solving tends to trigger a sympathetic response.

External attention, or the intent to "note and detect external stimuli" is found to result in cardiac deceleration, a parasympathetic response. According to Lacey (1967), "the decelerator seems to be wide open to his environment, keyed for the reception of input and for the release of simple responses to these inputs." Lacey (1967) also concludes that "attentive observation of the external environment...can modify the somatic response to an intellectually stressful problem-solving task." This means that one's intent, attitude and approach to a task are powerful factors in determining the physiological and biochemical responses that occur.

Easterbrook (1959) hypothesized that high arousal results in a narrowing of attention from secondary peripheral information toward the central information coming from a primary task. While Easterbrook considered this narrowing of attention to be a relatively passive and automatic process. Eysenck (1982) proposed that it could be viewed as an active coping response. Eysenck also sees anxiety as being a major factor leading to a narrowing of attention concomitantly resulting in the tuning out of peripheral information.

Kahneman (1973) relates effort to arousal. He sees increased effort as leading to increased arousal. In addition, he feels that attentional capacity or the energy and channels available for performing mental operations is also related to arousal. In the early stages of an attentional situation (low arousal) such as a task involving analysis, the attentional capacity seems to increase with increasing arousal. At high levels of arousal, a reduction in functional attentional capacity occurs (Hasker & Zacks 1979). In this way, Kahneman equates arousal with both effort and attention.

Different heart rate responses also appear with different problem-solving tasks that one is engaged in as well as with how one approaches the task. It has been noted that in categorizing tasks, the heart rate tends to decelerate. On a picture-matching categorizing task involving young children, there was a marked heart rate deceleration until the choice was made. The greatest deceleration was related to making a correct response (Lewis & Wilson 1970). In reasoning tasks, the heart rate tends to accelerate. This has perplexed many researchers who have tried, often with less than satisfactory results, to reconcile the apparent enigma. Both tasks obviously involve a degree of problem solving, attention, effort and arousal. What then, is the difference?

According to Lacey's approach, categorizing involves more "looking" and being more wide open to the environment, to the periphery and to secondary elements in the surround, a state that triggers cardiac deceleration. Reasoning involves more internalization and a reduction in external awareness, a state that would trigger cardiac acceleration.

Pribram (1975) comes to a similar conclusion but does so via a different route. Remember that some investigators, such as Kahneman, equate arousal with effort and attention. Others equate arousal with activation. Pribram considers attention, arousal, activation and effort to be separate entities. Arousal, to Pribram, is indicative of only one kind of attention, activation of another and effort comes to the fore only when the situation calls for some coordination or regulation to take place between the two in order to produce a change in the status of information processing.

In the case of categorizing versus reasoning, Pribram offers the hypothesis that in categorizing, arousal (the intake of information) precedes activation (the preparation for action). In reasoning, the opposite occurs. Activation precedes arousal. It is this distinction that results in the different types of heart rate responses to these tasks.

Pribram also makes a second distinction. He notes that in passive attention, where one accepts maximum external information, the heart rate decreases. In situations where one rejects extraneous information, the heart rate increases. This is quite similar to Lacey's hypothesis. The question is how this relates to the notion of arousal preceding activation or activation preceding arousal.

To Pribram, arousal represents the "What is it?" reaction. In situations where arousal precedes activation, it is inferred that a priority is given to being more attuned to what is "out there" and what it may be rather than to performing actions on it. In other words, more attention is being given to the information itself.

In situations where activation precedes arousal, it is inferred that a priority is being given to acting on the information. More concern is placed on the readiness for doing something than just in recognizing what is out there. Remember that activation, to Pribram, represents a readiness to respond, to be more in tune to "What's to be done." Pribram uses the example of playing chess where one makes calculations and plans out moves and then, only after the move, does one observe whether it was actually the right thing to do or not. In this instance, the preparation for response (activation) precedes observing the results (arousal).

In either case, it is not an all or nothing situation. Most tasks are never approached with total arousal and no activation or total activation and no arousal, at least in Pribram's use of the terms. Both aspects operate simultaneously in conjunction with each other. There are elements of each involved in any particular task and at any particular moment. It is all a question of priority. And assigning priority is influenced by the nature of the task and by the nature of the individual. Some tasks require more emphasis on what something is than doing something about it, and different individuals prefer, or are more comfortable with, one operational mode over the other. Either they or the task may be more arousal-oriented or more activation-oriented.

A preference toward being arousal-oriented so that arousal takes precedence over activation implies being more open to input, more open to all varieties of potential information and more open to the periphery. As noted before, the nature of a categorizing task involves being more receptive to what is there, being more alert to the total environment and more open sensorially. It is this that appears to result in cardiac deceleration.

A preference toward being activation-oriented so that activation takes precedence over arousal implies being more predisposed to action, to "go and do." This entails being primed either to attack or flee. Both involve motoric responses. And, the act of reasoning involves a motoric vigilant readiness that results in muscular tonicity. According to Pribram, in categorizing there is a decrease in isotonic contraction while in reasoning there is an increase. As the intensity of concentration increases, as muscular tonicity increases, intake of information (arousal) decreases which appears to be the trigger for cardiac acceleration.

In Pribram's sense, arousal has to do with being perceptually responsive, being open to all sorts of stimuli and being sensitive to novel input of all kinds. This increases the degree of entropy or uncertainty in the system. Activation, on the other hand, is more internally regulated, involving controlled vigilance rather than simple perceptual responsiveness. This control provides a bias against change in the output channels. It emphasizes routinization, order and redundancy which lead to "continuity and stability in motor control" (Tucker & Williamson 1984). This may be another factor that triggers differences in heart action: the difference between entropy (openness to change) and redundancy (the need for control) with the latter causing a constriction of perceptual attention that allows the organism to narrow its focus to minimize distraction during what it considers to be adaptively important situations. Unfortunately, what was adaptively important for primitive man in terms of basic survival has been superseded by more achievement-oriented goals in modern civilized man.

Arousal and Stress

Although the elaborate Pribram model with its specific and narrow definitions of arousal and activation has been received with much interest and some support, the use of the term arousal in the general literature still relates to "being aroused." Its definition has become so broad that, depending on the particular researcher, theorist or writer, it may or may not be used to represent factors such as awareness, attention, alerting, orienting, motor responsiveness, effort and activation. Its usage is similar to how Selye broadened the stress concept so that all activities, events, interactions and even thoughts are considered to be stressors. It is also similar to how the general use of the term stress often encompasses the stimulus, the response to the stimulus and being anxious or distraught.

In practical usage, the terms arousal and stress have broadened to include so much that is similar that a point has been reached where it is difficult to discern any difference between them. After all, things considered as being equal to the same things must be considered as being equal to each other. Whether this is totally or partially true in this instance is another question.

One thing that has occurred is that the distinctions between arousal and stress have become blurred.

Chapter 9

The Enigma of the Pupillary Response

Eckhard Hess (1964) opened up a new area of research for both arousal and stress investigators with his work on pupillary reactions. Actually, interest in the pupil and its responses is not new. Darwin (1872) related pupillary dilation in animals to fear. Claude Bernard related pupillary action to autonomic nervous system function. A German neurologist by the name of Bumke is quoted by Hess (1975a) as noting early in this century that "Every active intellectual process, every psychical effort, every exertion of attention, every active mental image, regardless of content..., particularly every affect just as truly produces pupil enlargement as does every sensory stimulus."

Bumke stressed pupil enlargement. This is an indication of sympathetic nervous system activation. Pupillary constriction is an indication of parasympathetic activation. Eckhard Hess (1972) found that pupil size tends to increase with mental effort, the increase being in proportion to the difficulty of the problem (Hess 1964). Kahneman and Beatty (1966) found that pupil size increases with "loading" (storing a series in memory) and decreases with "unloading" (repeating the series aloud) with maximum dilation occurring just prior to responding. Gardner et al. (1975,1978) found that pupillary dilation is specific to mental encoding and retrieval of information rather than merely to mental effort as Hess noted. Beatty and Kahneman (1966) determined that pupillary dilation occurs relatively fast, within one second after the presentation of a demanding stimulus. In addition, Kahneman (1973) noted that the pupillary dilations which accompany correct responses in problem solving are usually larger in magnitude than those which accompany failures.

Besides its relationship to problem solving, pupil size has also been found to increase with anxiety and alertness, being slightly larger in those with high IQs than in those with low IQs (Janisse 1977). Pupil size has also been related to attitudes and emotions. For example, pupillary dilation is seen to occur when subjects taste something pleasureful as well as when viewing a pleasureful image, a preferred political figure or sexually stimulating material (Hess & Polt 1964, Hess 1965). Pupillary dilation is also observed when a heterosexual individual views dilated pupils in someone of the opposite sex and when a homosexual individual views dilated pupils in someone of the same sex (Hess 1975b).

It is well known that the pupil responds almost reflexly to a number of situations (Tryon 1975, Andreassi 1980). Changes in overall illumination result in changes in pupil size (light reflex). A change in illumination to only one eye will affect pupillary size in both eyes (consensual reflex). Pupil size is affected by accommodation (near reflex), blinking (blink reflex), by a constant fluctuation called hippus (pupillary unrest), fatigue, age, alcohol, drugs and medications. Pupillary responses are also associated with central nervous system and peripheral nervous system activation (Janisse 1977, Beatty & Wagoner 1978).

Although pupil size is influenced by all these factors, the measurement of pupillary dilation during problem solving, loading, mental effort, deep empathy, etc., and its return to normalcy only when the task is completed, has been proven to be consistent and repeatable. The action of the pupil in arousal, stress or attention happens to be a very sensitive measure of autonomic nervous system activity, mainly sympathetic activation with sympathetic reflex dilation accounting for from four-fifths to nine-tenths of the total extent of dilation. Less than one- fifth of the total dilation is found to be due to third nerve influences (Lowenstein & Loenfeld 1950). This consistency has made pupillometric measurements valuable in understanding the physiological responsiveness of neurotics and psychotics.

It has long been noted that neurotics and psychotics exhibit differences in autonomic nervous system reactivity as compared to normal individuals (Hoskins 1946, Eysenck 1956). In the normal individual, the expected response to a stressor is an increase in sympathetic-adrenergic outflow and a decrease in parasympathetic-cholinergic outflow. To determine if neurotics and psychotics (schizophrenics) exhibit the same stress response as do so-called normals, Leonard Rubin (1974) utilized a "cold-pressor" test as a stressor. This is a test in which a subject's hands are placed in ice water for ten seconds. Pupillometric measurements were then taken under both light- and dark-adapted conditions as a measure of the physiological response.

As expected, Rubin found that every normal individual, whether light- or dark-adapted, reacted to the stressor with a dilation of the pupil. Immediately after the termination of the stressor, there was a decrease in the pupil size of those with both light- and dark-adapted pupils.

Neurotic individuals were found to respond exactly as did the normal subjects to both light and dark changes as well as to the stressor. The only difference was that the neurotic individuals took longer to recover from the stress. It appears that parasympathetic activation did not come to the fore as it should have when the stressor was terminated.

On the other hand psychotic, or schizophrenic, subjects did not demonstrate a normal response either to light and dark changes at rest or under stress. This variation from the normal was so consistent that Rubin was led to conclude that if anyone "manifests aberrant pupillary constriction or dilation at rest, the overwhelming likelihood is that he is psychotic."

Now we have to pose some questions. Does this supposedly consistent pupillary response represent a more generalized systemic sympathetic response or is it strictly localized to the pupil of the eye? Is the sympathetic pupillary response to a cognitive task a physiological reflection of Cannon's "fight or flight" response? Is it equivalent to Selye's stress response? Remember that outside of a conditioned stress response, a so- called maladaptation, Selye postulates an almost immediate attempt at equilibration. What then does the sympathetic pupillary response really represent? And, does it support the hypothesis that arousal, attention and stress are all the same or at least are similar in nature?

One thing that should be kept in mind is that many different functions may utilize similar mechanisms. For example, tasting, chewing, talking, whistling, singing, kissing, lip-wetting and grimacing all involve a certain amount of tongue action. Just because tongue function is common to all these activities does not necessarily mean that these activities are all the same. Not every measure of high intraocular pressure is necessarily glaucoma. Not every high blood sugar finding necessarily represents the same form of diabetes. Is it possible, therefore, that an apparent sympathetic pupillary response might represent something other than stress, "fight or flight" or arousal?

The work of many researchers has repeatedly indicated that problem solving causes an increase in pupil size of from 0.1 to 0.5 mm, which is maintained until the problem solving is finished. How does this reconcile with cardiac deceleration during categorizing tasks? Is it possible to have a sympathetic pupillary response and a parasympathetic cardiac response at the same time?

The very fact that psychotics show varied and atypical pupillary responses to a stressor indicates that general trends are not necessarily universal trends. Conditioned sympathatonic or parasympathatonic adaptive responses further indicate that there can be variations from the normal. In spite of these variations, there are general trends of autonomic nervous system function that are noted in the majority of individuals who have been monitored physiologically. These can be divided into physiological and behavioral effects (see Table 2).

Table 2.

Physiological and Behavioral Effects of Autonomic Nervous System Function

Sympathetic Dominance	Physiological Effects	Parasympathetic Dominance
	cardiac rate	
	blood pressure	
	sweat gland secretion	
	blood glucose	
	fatty acid production	
	skeletal muscle tone	
	gastrointestinal function	
	respiration	
	pupil size	
	hormonal secretions (e.g., ACTH, adrenaline, thyroxin)	
	Behavioral Effects	
	arousal	
	attention	
	alertness	
	directed action	
	emotional responsiveness	

Physiologically, sympathetic activation generally produces an increase in heart rate, blood pressure, sweat gland secretion, breathing rate, blood glucose, fatty acids and in pupillary size. At the same time, there is a decrease in gastrointestinal function. Parasympathetic activation generally produces the opposite effect: a decrease in heart rate, blood pressure, sweat gland secretion, breathing rate, blood glucose, fatty acids and pupillary size while mediating an increase in gastrointestinal function.

Behaviorally, sympathetic activation generally produces an increase in arousal, alertness and emotional and intellectual responsiveness while parasympathetic activation results in a decrease in these responses. Again, since these trends appear to be so consistent, does pupillary enlargement and cardiac deceleration during a categorizing task indicate that it is possible for different autonomic nervous system responses to occur at the same time in response to given situations?

Another, more important question is how does the pupillary response relate to the accommodative (focusing) response? Should a sympathetic pupillary enlargement be expected to occur simultaneously with a sympathetic relaxation of accommodation? Perhaps looking into the subject of retinoscopy will help in answering these questions.

Accommodation and the Pupillary Response

Retinoscopy is a technique that has been used for over a century as an objective means for determining the refractive status of the eye. Variations in retinoscopic findings found during an examination generally have been attributed to changes in accommodation. Over the years a number of retinoscopic techniques have been devised to probe accommodative responses especially at nearpoint. These include bell retinoscopy (Pomeranz 1964) and stress-point retinoscopy (Kraskin 1965). A more important procedure, in terms of relating accommodative changes to cognitive demand, is one originated by Gerald N. Getman, O.D., and known as book retinoscopy (Pheiffer 1955, Getman & Kephart 1958, Getman 1960, Saltysiak 1969). A modification of this technique designed to minimize error in measurement is called the Monocular Estimation Method (MEM) (Haynes 1960, Bieber 1974). However,in its attempt to control sources of error, MEM becomes a decidedly different test from book retinoscopy.

Book retinoscopy basically involves the observation of changes in the retinoscopic reflex, mainly brightness, color and motion, while the patient reads graded material. In the case of preschoolers, it could involve viewing pictures or toys held at the plane of the retinoscope. Interest in the material, increased attention, the search for meaning-getting and the desire to problem-solve have been associated with a shift toward less "with" motion and greater "against" motion as well as a brightening of the retinoscopic reflex. "With" motion represents a decrease in accommodation, a tendency to focus behind or beyond the plane of regard. "Against" motion represents an increase in accommodation or a tendency to focus closer than the plane of regard.

Behavioral optometrists have also noted other relationships from the clinical use of book retinoscopy. For example, better performers on a reading task are not usually seen to lock accommodation rigidly onto the target they are viewing. They tend to exhibit a fluctuation in the retinoscopic reflex from a base- line posture that is slightly behind the task ("with" motion), a sympathetic nervous system response. When a point of difficulty is reached, there is a tendency to shift into an "against" motion or increased accommodation, a parasympathetic response. The greater and more prolonged the difficulty, the greater and more prolonged the cognitive task and the desire to pursue it, the greater is the tendency to posture accommodation closer than the plane of regard.

On the other hand, poorer achievers, or at least those who do not try to achieve, who daydream or for whom the task is too far above their comprehension level, tend to show a greater and duller "with" motion indicating more of a sympathetic response as they de-tune from the task at hand.

In a more structured research study, using an infrared recording retinoscope, Kruger (1980) found an increase in accommodation as the task demand was increased from reading numbers to actively adding them. In an interesting variation, Malmstrom et al. (1980, 1984) noted that increased mental concentration resulted in a tendency for accommodation to decrease, to shift away from the plane of regard toward the distance. The mental activity given to their subjects consisted of looking at a non-meaningful visual target (an "X") and keeping it in focus while doing a non-related cognitive task such as counting backward, starting from a given number and writing the numbers down.

Rather than being contradictory, these studies actually lend support to the basic clinical observations of book retinoscopy. "With" motion (decreased accommodation) or posturing accommodation beyond the plane of regard occurs with a less intense demand and especially when one's cognitive state is less related to the task at hand. The more intense the demand, the greater the effort being made to problem-solve, the more related the cognitive aspects are to the visual task, the less open to other aspects of the environment, the greater will be the tendency for "against" motion to result. As a matter of record, Kruger's task demands were far more intense, far more related to the specific visual task and involved far more mental effort and cognitive demand than those of the Malmstrom study. This alone could account for the different results each obtained.

The clinical observations from book retinoscopy, then, suggest that a slight base-line sympathetic accommodative response ("with" motion) may be essential for maximum efficiency on visual and cognitive tasks. They also suggest that a greater sympathetic response (increased "with" motion) may be an indicator of not being totally tuned into the ongoing visual task.

In answer to one of our previous questions as to whether it is possible that a sympathetic response might represent something other than stress, "fight or flight," or arousal--the answer appears to be yes. An observed sympathetic response may indeed be an indication of an immediate transient stress reaction as Selye predicts. It may also be part of a chronic conditioned sympathatonic response. At the same time, an observed sympathetic response may also be part of a mechanism for different types of functional actions. The assumption cannot be made that an observed sympathetic response is always related to either stress or arousal. It just may not be.

There is another question, however, that is still unanswered. Book retinoscopy, as we have seen, demonstrates that a subject striving to resolve an increased problem-solving demand tends to exhibit a parasympathetic accommodative response ("against" motion) at the exact time that pupillometrists note a repeated and consistent sympathetic pupillary response. Therefore, is it possible to have a sympathetic pupillary response and a parasympathetic accommodative response simultaneously? Much rides on this question.

It is a known fact that different physiological systems can react differently to a given stressor. Because of this, some researchers feel that what is generally called arousal is not a single process but a composite of many different functions that may or may not be going on simultaneously. There are those who already consider that what is measured as electrophysiological arousal, autonomic arousal and behavioral arousal are not different measures of the same thing but completely different organismic activities (Lacey 1967).

One organ or group of organs may also respond sympathetically to a given stressor while another responds parasympathetically to the same stressor. Could it be that this is what happens in the case of the pupillary and accommodative responses? After all, the pupil responds to light without correspondingly affecting accommodation. It is also possible to use drugs such as miotics and mydriatics to change pupil size with a minimal effect on accommodation. It would appear that a mechanism might exist for these two functions to operate somewhat independently.

Most research papers or texts on the subject of stress and arousal acknowledge pupillary action but have little or nothing to say about the subject of accommodation as it relates to the pupillary response. Some, like Hess and Polt (1964), try to eliminate accommodation as an interfering factor from their study. Others acknowledge that though it is theoretically possible for the pupil to respond differently than accommodation, under the effects of stress or arousal, however, it is highly improbable that it would do so.

Kruger (1977), noting that an increase in the brightness of the retinoscopic reflex occurs with an increase in pupil size and/or an increase in accommodation, offers a possible compromise by conjecturing that a sequential action might be involved. He suggests that a slight dilation of the pupil and relaxation of accommodation is followed by a slight increase in accommodation with a slight constriction of the pupil. He proposes that the initial enlargement of the pupil would act to reduce the depth of focus. This would be followed by a slight increase in accommodation of no more than 0.25 to 0.50 diopter in order to reduce the lag of accommodation. This slight increase in accommodation would result in the "against" motion seen during retinoscopy while the resulting slight decrease in pupil size which it would cause might be too small to be noticed. The only difficulty with this view is that this secondary slight decrease in pupil size has not been reported by those involved in pupillometric research.

Some recent research on accommodation in cats (Lovasik & Beauchamp 1982a,1982b) has revealed the apparent existence of accommodation neurons in the reticular formation of the brain stem not previously associated with accommodation. This is not only an unexpected finding, it also establishes a possible association between accommodation and both arousal and attentional mechanisms, at least in cats. In addition, and perhaps of even more importance, pupillary dilation has been observed in the presence of both positive accommodation and convergence (Lovasik & Beauchamp 1982a, Elul & Machiafava 1964, Hughes 1972). This runs counter to what one traditionally would expect (Davson 1976).

Is it possible that something similar may occur in humans? Though the physiology of the pupillary and accommodative responses are slightly different in cats as compared to humans, much insight into human function has been gained in recent years through research with laboratory animals. It would also seem logical that pupillary dilation in the presence of both cardiac deceleration and retinoscopic "against" motion would be a strong indication that this opposing type of functional relationship might, in fact, be possible in humans.

Arousal and Stress: A New View

Based on all the factors that we have been discussing, it may be useful to attempt to reconcile some of the enigmatic elements.

First, the evidence appears to indicate that arousal, in its traditional sense, and stress are far from being different names for the same act. As a general operating framework, it may be useful to consider arousal as a basic orientation either to the environment in general, as in awakening, or to some particular aspect of the environment. Stress, in Selye's broad sense, can be considered as the organism's actual response to this awareness, with distress or negative stress resulting when the organism's response is not in a corresponding relationship with the stimulus or when the interaction has an adverse effect on performance. This, incidentally, does not conflict with Pribram's interpretation of activation as being the "readiness to respond."

Second, as there is evidence that indicates that it is possible for there to be a different autonomic nervous system response to the pupil than there is to accommodation, it may be useful to consider that the pupillary and accommodative responses may often represent the action of different functional processes. The pupillary response, for example, can be considered related to intent, or how one is primed to respond to a stimulus or stimulus event. It appears to represent arousal in the sense that it demonstrates a basic interaction or involvement with a stimulus. At a more fundamental level, however, it can be considered to represent the intention that is necessary to generate or stimulate attention.

Outside of the act of awakening from sleep, arousal in this sense depends to a great extent on the choice of the individual to attend. A sheet of printed paper placed before an individual does not in itself create arousal. Arousal occurs only when that paper is reacted to with interest, possibly because of its novelty, possibly as a threat, or possibly as something to be read. It represents an attachment of sorts or at least a basic relationship between oneself and the stimulus. It is this "interest" that is demonstrated by the sympathetic pupillary response.

On the other hand, the accommodative response, as well as the cardiac response, can be considered as representing not the intent, but the coping that goes on in response to the stimulus. It indicates the actual ongoing "content" of the interaction, its quality, intensity and degree. The accommodative response is the mode and degree of the attention that stems from the intention. It is how one organizes or reorganizes one's central-peripheral (figure-ground) function in order to solve the problem at hand.

The approach that is being suggested is that the pupillary response can be viewed as indicating the individual's state of arousal while the accommodative reaction can be viewed as representing a measure of the individual's stress response.

There are other implications, some far more clinically relevant, that relate to visual function. These we will come to later. At least, at this point, we can see how some knowledge of the concepts involved in stress and arousal theory can aid in giving us a better understanding of what might be going on in the visual system.

Chapter 10

Burnout

Let us review a few factors about stress in general. First, regarding Selye's overall view of stress: Stress, in Selye's terms, is a response to a stimulus. It is characterized by a syndrome of common signs and symptoms. Although some researchers differ, the stress response to Selye is non-specific. This means that any stimulus can elicit it.

Second, in terms of the stressor: Since the stress response is "non-specific," all experiences, events, reactions and interactions are stressors of some sort to the organism. Some stressors, however, are positive, necessary, and supportive of life. These produce eustress. Other stressors are potentially destructive. These produce distress. The particular stressor is less important than its intensity. Extreme joy can be as stressful as extreme sorrow. The least stressful interaction is one accompanied by indifference.

The intensity of the stressor is dependent on a number of key factors. One of these is the figure-ground aspects of attention, input overload, and input underload. Another is the effect of one's personality and attitude. These influence how we perceive and interpret a particular experience as well as how we react to it as a stressor.

Stressors can be general, acting on many parts of the body at the same time, or local, affecting only one part of the body. They may be single or multiple. A general stress can affect a local stress. A local stress can influence a general stress. And, numerous local stressors, or a single intense local stress, can trigger a general stress.

Third, regarding the stress syndrome: The General Adaptation Syndrome (GAS) and its local counterpart, the Local Adaptation Syndrome (LAS) are Selye's paradigms for how the organism reacts to stressors. The syndrome is composed of three stages: the Alarm Reaction, the Stage of Resistance, and the Stage of Exhaustion.

In most instances, the neural and endocrine systems that are initially activated by the stressor are sympathetic-adrenergic. Parasympathetic-cholinergic activation is generally used to restore normalcy.

The stress response can be conditioned. It can become sympathatonic, in which case the initial alarm stage is perpetuated. It can become parasympathatonic in which case the initial alarm stage is often bypassed while the adaptive response is grossly exaggerated. Either of these reactions are maladaptive responses producing what are called "diseases of adaptation."

We have seen, therefore, that stress or the stress response is normal to everyday life, paralleling it and representing, according to Selye (1976),"the wear and tear of the body caused by life at any one time."

We have also seen that all stressor agents provoke a physiological reaction of some sort. Many tend to provoke a psychological reaction. Of even more importance is the fact that one's psychological, emotional, or attitudinal "pre-set" strongly affects the stress response. This simply cannot be emphasized enough.

There have been many experiments that have demonstrated that it is the expectation and not the objective condition that determines how we interpret what we interact with and how we interpret ourselves. Glass and Singer (1972) have pointed out that exposure to stressors that are perceived as being unpredictable and uncontrollable not only can produce an increased initial stress response but can induce negative performance effects after the stimulation is over, during and after adaptation. This means that the perception of a stressor as being unpredictable and uncontrollable leaves one less able to cope with subsequent demands and frustrations. Regaining the feeling of being in control and of being able to predict outcomes ameliorates these effects. As Tubesing (1982) has concluded, stress is not inherent in the event itself, but in how one perceives that event. Change one's perception of an event and you automatically change the stress reaction to that event. Both high self-esteem as well as depression often come from our interpretation of how we affect the world. In fact our entire relationship to the world around us is based on our cognitive set, biases and belief systems.

Therefore, when we say that a task is demanding or that it is stressful we are actually referring to two major interacting factors: the stressor itself and the psychological approach to the stressor. A stressor can be intense or prolonged but it is one's interpretation of the stressor, how it is perceived, and one's general pre-set to life itself that determines how demanding that particular task really is.

The best illustration of this is "burnout." Burnout is the end result of repeated and/or prolonged intense stress. To be more accurate, it is the extreme result of what is perceived to be repeated and/or prolonged negative stress, or distress. Burnout is a breakdown, a collapse of function that is directly related to the perception that one is no longer able to cope, to the perception that one is a victim of an unfriendly, uncooperative, unpredictable and uncontrollable universe. It is a sense of failure, a loss of hope in the ability to realize one's ideals. In burnout, one simply declares oneself to be "out of business" (Morrow 1981).

Burnout happens to be a mushrooming symptom of our modern society. Pressure has become part of everybody's life--in school, at work, in sports, and at home. It is fostered by the need to succeed, achieve, compete, gain more knowledge, get things done faster, and by simply having to wait. Everyone has to wait for appointments, for gas, to check out in the supermarket, to make a bank transaction, to pay a toll, to be served, for the telephone to be available, to board a bus or plane, etc. In addition, pressure is present when we overly concern ourselves with health or disease. It is present when we are preoccupied with stress. It is present when one's performance is inadequately appraised and inadequately rewarded. Pressure is aggravated when we measure ourselves against others and perceive ourselves as being inferior. It is aggravated further when we reject some part of ourselves or some dimension of our life (Swogger 1981) rather than viewing life and one's participation in it as making sense and as being coherent (Antonovsky 1974).

It is hard to determine when stress becomes burnout. Most people are not usually aware of when they are tense or anxious. Yet there are physical and behavioral signs and symptoms that reveal the presence of the type of tension that leads to burnout (Morrow 1981, Swogger 1981, Antonovsky 1974, Hensley 1984, Belcastro & Gold 1983, Maslach 1976, Veninga & Spradley 1981). In fact,the more numerous and the more intense are these signs, the more is the chance that the individual is approaching or is already into that state (see Table 3).

Burnout is a very serious condition because besides unhappiness, exhaustion and loss of efficiency, there is always the chance that the burned-out individual may utilize the ultimate escape from his perceived problems--suicide, a final statement affirming that all his irrational, magical beliefs are real to him (Serban 1982).

It is interesting how many of the signs and symptoms of burnout are simply extremes of flight or fight. In most instances the flight component is really a form of fight or attack on others as well as on one's self. Any decisive behavioral action is usually a symbol or a metaphor for an underlying statement. And, a decision to flee or avoid a perceived situation as opposed to a reflex withdrawal to an actual threat, is, in essence, an aggressive act.

Table 3

Typical Signs of Burnout

Physical Signs	
insomnia	fatigue
palpitations	nausea
headaches	loss of appetite
shortness of breath	weight loss
nervousness	ulcers
trembling	hypertension
exhaustion	lingering colds

Behavioral Signs	
feelings of discontent	extreme suspiciousness
loss of concern	extreme frustration
irritability	periods of extreme cynicism
overreaction to criticism	a decline in performance
outbursts of anger and hostility	a lack of enthusiasm for one's work
forgetfulness	inability to make decisions
making snap and usually negative judgements, especially about people	excessive daydreaming
	excessive drinking, smoking and drug usage
avoiding being with people	feelings of being powerless to control one's life
disappointment in others	
feelings of failure	a feeling that one is at the end of one's rope
feelings of despair	
feelings of guilt	feelings of helplessness and hopelessness
depression	
lowered self-esteem	
feelings of alienation	
apathy	
feelings of being overloaded	

Burnout is unique in that it has less to do with a person's interaction with things than his interaction with others. It is people-oriented. It also relates to personal power or at least the perception of being able to exert control over given situations. In addition, there are also burnout-prone individuals. Maslach (1982) lists three sets of traits that predispose one to burnout: (1) being weak, anxious, submissive, unassertive and fearful of involvement, (2) being impatient, intolerant and easily angered and frustrated by obstacles in one's path, and (3) lacking self-confidence, having little ambition and being more reserved and conventional.

There are obviously many factors for becoming a candidate for burnout. One relates to an individual's personality. A second has to do with the situations that one is placed in. Some children, parents, employers, associates and even friends are just more difficult to get along with than others. Some home, school or work situations are exceptionally demanding. Economic factors also play a part. Yet, with or without these factors, more and more people are demonstrating greater feelings of vulnerability. There are many possible reasons for this. One has to do with the effect of technology.

It is difficult not to be impressed with the technological age we live in and how we seem to be mastering both space and time. Communication by sight and sound is becoming almost instantaneous regardless of the distance involved. Most of us have become used to telephones, television, word processors, computers, telescopes, microscopes, electricity, automobiles, bicycles, trains, buses, airplanes, space craft, air conditioning, sunglasses, thermal underwear, etc. It also appears that almost everything one needs is practically at one's fingertips, or at least is potentially available except in severely economically- deprived areas. However, although economically and culturally deprived individuals may also have reason to become burned-out, it usually happens for different reasons since they have many things going for them that more endowed communities have lost. Underneath the impressive cultural facade of modern civilization, the individual has become more helpless, more dependent on society and more dependent on the talents of others who are, by and large, anonymous.

Just consider how many individuals today could manage or survive if left by themselves on the proverbial desert island without the conveniences we take so much for granted. How many would know how to improvise shelters, make clothes, and obtain and cook edible, nonpoisonous food? How many would be able to make a fire without special lighting devices? How many could make shoes, shoelaces, belts, combs, tooth brushes or eating utensils? How many would know how to make furniture, cultivate produce, domesticate wild animals and care for farm animals? How many could care for their dental and medical needs?

It was not too long ago when people could take care of most of their own basic needs and those of their family. They built homes, farmed, hunted and cooked. If more expert help or simply another pair of hands was needed, it usually came from close neighbors or family. Outside of major accidents, war, disease or natural catastrophes, these people could at least feel that they were in partial control of their lives. Today, however, too many realize their almost total dependence on forces outside themselves. They rely on society, teachers, specialists, government, parents, employers, physicians and even computers to do things for them or at least to tell them what to do. They rely on the technological advances of society to take care of their basic needs. This only adds to their feelings of insecurity, vulnerability and helplessness. Deep down, many individuals sense that others are really in control, at more and more levels, of their lives.

People's actions and thoughts today reveal the result of this insecurity. More are saying, in effect, "I want to be in control," "I want to be heard," "I want my vote to count," "I want to be myself," "I want to do my own thing" while inwardly they feel a sense of frustration as their inner voice retorts, "But you are only a little cog in a big wheel," "You can't buck the system," "How can you be in control when you don't know how to be in control. You never learned. Everything has always been done for you or to you. You have always conformed."

Then, in attempting to create the illusion that they are in control, more and more individuals of all ages begin to attack others in some way. They become aggressive, demanding, insulting and pushy. They strive harder to achieve. Their actions demonstrate the statements, "See, I am somebody," "I can be something," "I can do something," "I really do have some control over events."

Some go to the other extreme by quitting their jobs and possibly even leaving their spouses and families. In this way, even "flight" is a personal choice that demonstrates some control over events. Yet in both instances, "fight" and "flight" are still forms of aggressive action as one attempts to reveal to oneself and to others that in some way one is in control. Of course, the need to do this stems from the inner feeling that one is not really in control. And, this pervading sense of not being in control is the factor that generates the feelings of vulnerability and helplessness that culminates in what has become known as burnout.

Another factor that precipitates or leads to burnout is the cultural difference between ourselves and our ancestor "Og." Few today are confronted with the same type of life-threatening situations as was Og. There are not too many sabre-toothed tigers around any more, neither in the cities, the suburbs or on the farms. The basic "fight or flight" response appears to have been designed as a physiological means of coping with actual, sudden, life-threatening situations. In such a situation, one mobilizes oneself either to fight back or run for one's life. Today, we make threatening situations out of examinations, a visit to the dentist, getting somewhere on time, doing a task properly, finishing a job or test on time, gaining approval and feeling the need to be somebody. As one becomes conditioned to respond maximally to a perceived threat rather than an actual threat, one begins to lose the ability to respond appropriately either to actual life-threatening situations or to relatively minor events. There is a tendency to overreact instead to all conditions. This only further increases the feeling of vulnerability. One also feels less in control because one's emotional, physiological and biochemical reactions almost seem to have a life of their own. People who ask themselves, "Why do I get so emotional?" "Why do I overreact?" "Why am I so anxious over nothing?" reveal loudly and clearly that their feelings seem to control them rather than the other way around. Overreacting is another way of being out of control.

The organism also reveals a basic physiological responsiveness to action of all kinds. The need to move and to act in and on the environment appears to be part of an inborn biological drive. Another precipitating factor leading to burnout is the conflict set up between this biological drive for physical action and society's stress on restraining action. Society encourages all forms of immobility. These include studying, viewing television or video display terminals, watching movies, riding in cars, trains, buses and planes and working for prolonged periods in confined areas on restricted paper and pencil tasks. Society also encourages the withholding of emotions, the suppression of thoughts, the repression of likes, dislikes, hates, hopes and fears and the inhibition of laughter, crying and blushing. All this contributes to burnout by the fact that we are constantly being primed by nature for action yet encouraged by society not to move, not to express action.

Chapter 11

Redefining Stress

We have discussed the stress response from varying approaches. Still some questions surface. For example, is stress simply a "coming to terms" with life's interactions? Is there possibly a broader and more fundamental way of looking at this whole process to determine what stress and the stress response might really represent?

To try to answer these questions let us view the subject of stress from a slightly different perspective, a more philosophical one, and possibly redefine some of its elements (Forrest 1980a).

In the previous chapter the point was made that we are primed by nature for action while society in many ways indirectly encourages the restraint of movement. In a still earlier chapter, we touched on how immobilization, a component of input underload, was a stress producer. This whole subject of movement and restraint of movement has broad implications.

First, it is obvious that movement of some sort and to some degree pervades all aspects of nature and existence (Forrest 1969). This appears true in terms of the macroscopic, the microscopic, the animate and the inanimate. It is common to both matter and energy, although it may be difficult to visualize energy as movement. Existence itself implies some form of movement. In terms of observable nature, therefore, movement is fundamental. In fact one way of viewing Einstein's E=mc2 formula is as a demonstration of the relationship among movement (in terms of the speed of light), energy and mass.

A second factor concerning movement is that although there are movement patterns in nature that appear to be identical and repeatable, no two movement patterns are ever the same. They may appear to be identical but that is only an appearance. Lightning never strikes the same place in exactly the same way with the same form and shape twice in a row. Molecules never repeat their actions exactly even if their overall pattern is similar. A train riding on a given section of track never duplicates its exact speed for the same length of time with the same lateral and vertical sway movements. Even the earth never rotates twice around the sun in quite the same manner. We may like to think that the rotation of the earth is consistent but that only happens to be wishful thinking. Somehow we seem to have a need for such consistency. Psychologically, it is very reassuring.

There are two important elements, therefore, in regard to the fact that everything at all levels is in constant movement. One is that inaction is illusory. There can be no inaction. There can be more movement or less movement, always in relation to something else, but there cannot be no movement. What we tend to call inaction simply represents a relative limitation, restriction or inhibition of potential movement capabilities.

The second element is that movement or action, since it is never exactly the same in all respects, represents change. What appears to be a consistency of action then is simply a restraint, not a total elimination, of change.

When we view everything about us, including everything we do or do not do or that others do or do not do, it soon becomes obvious that all of life appears to involve an interplay between two factors: movement, action or change on the one hand, and apparent inaction, resistance or restraint of movement on the other hand.

Another way of viewing the same thing is through the concepts of entropy and redundancy (Shannon 1948,1956,1951; Von Neumann [1949] 1966; Campbell 1982). In its traditional sense, entropy represents the disorderliness of a system, its degree of changeability. The greater the entropy, the more disorder. The more disorder, the more uncertainty and the more unpredictability. The ultimate in entropy is chaos. One advantage of entropy is that the greater the disorder, unpredictability and uncertainty, the more potential ways there are of structuring the system.

Anything that counters entropy such as structure, order, rules or regulations are aspects of redundancy. Redundancy increases the repetitive nature of an action and makes it more predictable. Redundancy reduces entropy by restraint of action. The existence of gravitational forces are a form of restraint on a more cosmic level. Play activities designed for children are a way of structuring behavior rigidly enough to enhance specific types of learning yet loosely enough to encourage creativity and inventiveness. Genetic coding also fosters redundancy. Rules for grammar, spelling, reading and writing, in fact rules of any kind are all forms of redundancy. Again, where entropy represents change or changeability, redundancy represents restraint or limitation of action.

The interplay between these two elements pervades all areas of life. For example, the terms evolution and revolution, whether they are biological, social or political, are simply different names for change. One just happens to be a little slower than the other. Evolution is less unsettling because we tend to be less aware of most of the changes as they occur.

Similarly, adaptation of any kind, whether it be biological, social or political, simply represents the attempt to minimize change, to preserve the status quo, in the face of the stimulus to change. The autonomic nervous system demonstrates this duality. The sympathetic nervous system primes the organism to respond to change while the parasympathetic nervous system mediates restabilization or resistance to change.

Homeostasis implies equilibration to resist change and maintain some form of internal constancy. If the body were not confronted with change and were not itself susceptible to some degree of change, why would there be a need for equilibrating mechanisms at all?

Language is also part of the process of restraint. Language, as it is used to label, categorize and communicate is at the same time both a substitute for action and a limiter of action. The very act of verbally specifying an object, act or event delimits and isolates it. Language, in effect, freezes action.

Even the difference between sensation and perception reflects the interplay between change and restraint (Forrest 1981c). Sensation is pure experiencing. It represents immediacy, what is happening now. If life is change, then sensation represents awareness of change as it occurs. Perception, on the other hand, which is usually considered to be "sensation plus meaning," actually represents awareness that is modified, tempered, stabilized or "pinned down" by one's experience, by the memory of what was. Perception makes what "is" less novel by tying it in with what one has already experienced in the past.

Psychologically, as well, there is the tendency to resist change. We seem to yearn for cradle-to-grave predictability. We tend to clutch the known and fear what is unknown. We often tend to become apprehensive over totally novel situations. In fact, those who are most dissatisfied with aspects of their lives often seem to resist any shift that could lead to great contentment and well-being. Those who want certain changes, who desire to have or do something new, usually want the changes only on their own terms. They want control of the situation at all times.

This presents an intriguing counterpoint. We see that, on the one hand, life, in general, gives an injunction for action while, on the other hand, everyday living entails the harnessing of action, the establishment of some degree of control. Where nature thrives on mobility (action, change, freedom, evolution, immediacy, awareness and creativity), culture thrives on immobility (laws, rules, edicts, mores, norms, language and logic). A given life, therefore, involves an interplay between change or action and adaptation or the neutralizing of change and the control of action.

It is possible to take the liberty of conjecturing that this operational counterpoint is built into the human organism for a purpose. For one thing, it appears to aid performance and function by providing ground rules for the consensus reality we live in. It appears to provide the illusion of stability, coherency, constancy, certainty and predictability so that one may be able to cope better with a universe that in different ways often appears unstable, inconstant, incoherent, uncertain, improbable and unpredictable, where change is the only constant.

Another way of looking at this is that this counterpoint may help an individual literally create the illusion of stability until he or she is ready psychologically to break away from the need for such stability and to become more aware and accepting of what truly "is" and perhaps even recognize that stability and unity exist in change. Unfortunately, too many get stuck in the illusion.

Now, we can begin to redefine stress in these new terms.1 Stress can be considered as the attempt to control and be in control of all forms of action. In this view, stress is resistance to all forms of change. It is not something that leads to adaptation. As Selye pointed out, it is the adaptive response, the force that strives to stabilize.

Using this perspective as an operational framework, we can hypothesize three major types of stress: (1) primary or biological, (2) physiological, and (3) functional or psychological.

1. Primary or biological stress is simply the result of the basic neuro-biochemical-endocrine interplay needed to survive and operate in the normal environment. It is the response of the body as it attempts to maintain stability and normality in the presence of more common, average-type changes in the internal and external environment. It is the process by which the body maintains its internal constancy under normal conditions, as Claude Bernard viewed it, or the process of homeostasis in Cannon's original sense, or what Herrick called "biological regulation."

2. Physiological stress represents the response of the body to isolate, minimize, stabilize, limit and control the effects of trauma, disease, input overload,input underload, exercise, concentrated work, problem solving, etc. It is the response to stronger than normal stimuli or stressor agents. It can be thought of as "homeostasis PLUS." It involves the response to Cannon's "fight or flight" reaction and Selye's "Stage of Resistance."

3. Functional stress represents the stress created by how one perceives an event. It is one's psychological resistance to changing life events and to the feeling that outcomes cannot be to-

tally predicted or controlled. It is the resistance that stems from knowing that one is not fully in control of a situation that one desires to control. Note that Cannon referred to his "fight or flight" reaction as being a response to a perceived threat, not necessarily an actual one. In most instances a threat is not a threat unless it is perceived, evaluated or appraised as being a threat.

The whole issue of control is based on the belief that an individual can influence the effects of an event. In reality, it does not have to be actual control. It just has to be the feeling, the perception, that one is in control or has control (Thompson 1981). It is the meaning given to the situation by the individual that determines its stress effect.

What we call functional stress, then, is the psychological appraisal that modifies the physiological response. To Appley and Trumbull (1967) the actual "induction of stress depends on the mediation of some appraising, perceiving or interpreting mechanism." David Glass (1980) views stress as "the perception of threat to physical or psychological well-being and the perception that (the individual's) responses are inadequate to cope with it." And Glenn Swogger (1981) observes that "a life pattern characterized by stress involves a rejection or lack of acceptance of some part of ourselves or some dimension of life."

In 1979, Holmes and Rahe (1979) published their now famous "Social Readjustment Scale" in which specific life events are given a numerical rating based on their potential effect on the individual. The score that one gets by totaling up all the key events that occur in one's life during a given period becomes the individual's stress quotient. The events listed on this scale all involve change. These include death of a spouse, divorce, separation, loss of a job, a jail sentence, injury, marriage, vacation, etc. (see Table 4). Supposedly, the higher the score, the higher the stress effect on the individual and the greater the likelihood that he will experience related mental and physical problems. Holmes and Rahe have shown that the probability of this occurring is greatest if a score of 300 life- event change units is accumulated within a six-month period of time.

In a similar vein, Yamamoti (1979) determined what key experiences children perceived as being most stressful to them (see Table 5).

These types of tables, scales, ratings and quotients all deal with a change in the status quo of one's life and the relative stressfulness of the change. Practically speaking, however, they must be taken only as general guides as to how one might respond individually and within a particular cultural group. This is im-

Table 4.

Life-Event Change Scale

Event	Score
Death of spouse	100
Divorce	73
Marital separation	65
Jail term	63
Death of close family member	63
Personal injury or illness	53
Marriage	50
Fired at work	47
Marital reconciliation	45
Retirement	45
Change in family member's health	44
Pregnancy	40
Sex difficulties	39
Change in financial status	38
Change in occupation	36
Foreclosure of mortgage or loan	30
Change in responsibilities at work	29
Son or daughter leaving home	29
Trouble with in-laws	29
Outstanding personal achievement	28
Spouse begins or stops work	26
Begin or end school	26
Change in living conditions	25
Change in residence	20
Change in schools	20
Change in recreation	19
Change in social activities	18
Change in sleeping habits	16
Change in eating habits	15
Vacation	13
Minor violations of the law	11
Total Score	___

Reprinted, with permission, from Holmes and Rahe, "The Social Readjustment Scale," *Journal of Psychosomatic Research*, (1979).

Table 5.

Children's Ratings of the Stressfulness of Experiences

1.	Losing a parent
2.	Going blind
3.	Being retained in a grade
4.	Wetting in class
5.	Parental fights
6.	Caught in a theft
7.	Suspected of lying
8.	A poor report card
9.	Sent to the principal
10.	Having surgery

Reprinted, with permission, from K. Yamamoti, "Children's Ratings of the Stressfulness of Experiences," *Developmental Psychology* (1979).

portant because the magnitude of a given stress response is determined not simply by the specific change, per se, but by how each individual views that particular change by one's psychological resistance or lack of willingness to accept the situation as well as by the fear one might have of future change. In this respect, these types of tables can never be totally representative of what actually occurs to a given human being, certainly not from culture to culture, nor from person to person, nor from moment to moment.

We spoke earlier about predictability, escapability and controllability. Actually the key word was perceived--perceived predictability, perceived escapability and perceived controllability. When one analyzes each of these different conditions, however, it becomes apparent that the perception of predictability, the feeling that outcomes can be predicted, and the perception of escapability, the feeling that one is not trapped in a given situation, are all just aspects of the perception of controllability, the feeling that one is or is not in control of events.

It is the feeling of not being in control, that as much as we try, events seem to have their own way, that things just happen to us, that often leads to the feeling of helplessness (Seligman 1975, Garber & Seligman 1980). A persistent feeling of helplessness can often lead to a feeling of hopelessness. And, a persistent, pervading, feeling of hopelessness has been demonstrated as being a potential precursor to severe mental and physical disorders including depressive states as well as to possible premature death.

A more common, less severe result of the feeling of not being in control is chronic anxiety, more often referred to as inappropriate anxiety. Anxiety is so prevalent in our culture that it can be considered to have become the psychological equivalent of the common cold. Specifically, anxiety is the tendency to get upset easily over minor situations. It is getting angry and yelling at a waiter who serves a rare steak that was supposed to be well-done. It is getting "butterflies in the stomach" when being interviewed for a job. It is getting heart palpitations anticipating being called upon in class or becoming upset because one feels inappropriately dressed for a special occasion. Just picture this type of anxious individual having to stand in a slow-moving line or being awakened from a deep sleep by the telephone ringing at 2:00 A.M., or being in a last-minute rush to catch a plane, train or bus.

The intriguing fact about anxiety is that, more often than not, it is completely unrelated to the reality of the situation. The anxious individual literally creates mountains out of molehills, magnifying the importance of events way beyond their actual significance.

It is commonly observed that few people like to wait in lines whether it be at the post office, the supermarket, in banks or at the theatre. More anxious people feel that having to wait in line at all is an imposition, as if the system is penalizing them unjustifiably. Those who are less anxious may only feel imposed upon when someone tries to get ahead of them in line or when, after waiting for hours in a long line to get tickets for a special performance, are told that tickets are no longer available. Others seem to be able to take most occurrences philosophically.

Again, it is not the particular situation, but one's approach to the event that makes the difference. Occasional anxiety and its more chronic variant, inappropriate anxiety, are examples not of events or physiological reactions but of a state of mind. In other words, "pressure" may be part of a situation of physical or social constraint but anxiety is the expression of how the individual relates to what is happening.

There are certain attributes common to those who exhibit anxiety. Anxious individuals tend to judge themselves too harshly in relation to what they perceive as their role models. These role models could be individuals or rules that parents, schools or society promulgate. They also feel that things or events are directed at them personally, as if people or the system are out to get them.

Anxious people also perceive that their actions and emotional responses control them, rather than the other way around. They just seem to happen.

Finally, the anxious individual tends to worry a lot in advance, creating a state of fearfulness even when the situation is not life threatening and where physical harm is not even a probability. This is the individual who worries about future events which might or might not happen, possibly saying, "I know the bus will be late," "I know that the telephone call will bring bad news," "I know that my symptoms mean that I have a fatal illness," "I know that my boss will fire me," "I know that the dental drill will be painful," or "I just know this plane will crash."

This is the type of individual who keeps mentally reliving past events, so many of which should have been forgotten and done with, and then projecting them into the future. This is the type of individual who talks a lot to himself, who is self-judgmental, with all the "woulds," "shoulds," and "might-have-beens" all contributing to feelings of guilt. These individuals are full of what Shel Silverstein (1981) calls the "Whatifs":

Whatif I'm dumb in school?
Whatif they closed the swimming pool?
Whatif I get beat up?
Whatif there's poison in my cup?
Whatif I start to cry?
Whatif I get sick and die?...
Whatif! Whatif! Whatif!

Functional stress, therefore, stems mostly from one's reaction to imagined or fantasized dangers, threats and fears. It is simply resistance to life as it is now. It is literally going against what "is."

Chapter 12

Stress, Attitude and Disease--I

According to Selye (1976), all life events and interactions are stressors. To Holmes and Rahe (1979) and Yamamoti (1979) certain key life experiences or changes from the status quo appear to have more of an impact on the individual than do others. Lazarus (1981) found that so-called daily hassles such as disagreements, disappointments, accidents and unpleasant surprises have an even greater effect on one's moods and health than do major life events. In fact, Lazarus feels that the effects that major life events have on an individual occur mainly through the daily hassles that they provoke.

Selye (1979) considers that stress, regardless of the source, has four basic components; hyperstress as opposed to hypostress and eustress as opposed to distress. He sees the basic biological goal of human behavior as attempting to strike a balance between the equally destructive forces of hyperstress and hypostress and, at the same time, finding as much eustress as possible while minimizing distress (see Figure 20).

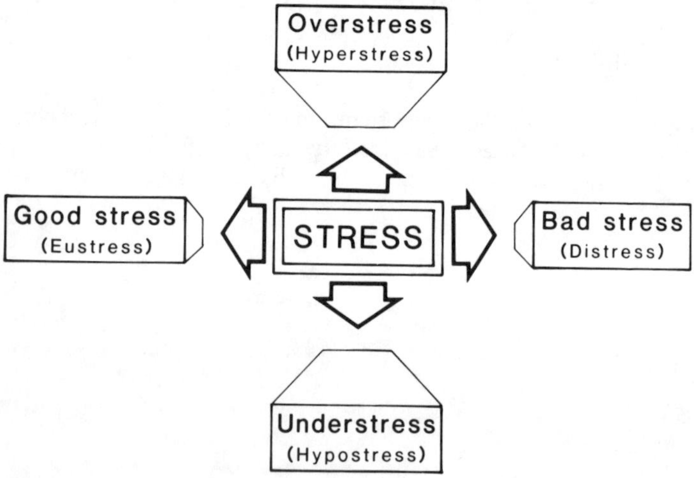

Fig. 20. Key components of the stress response

Selye (1974a) also recognizes and includes psychological influences, especially the element of frustration, as major variables affecting the type of stress reaction that occurs. He notes, however, that even the greatest experts in the field are not quite sure why frustration is much more likely to produce negative effects of distress than is excessive muscular effort. A possible answer to this is that, practically speaking, frustration is simply a description of a particular reaction to those many factors that hinder one's ability to cope and which trigger the feeling that there is no escape from a perceived impasse so that, in effect, one is not in total control of the situation at hand.

All the psychological factors, including frustration, cast their influence in two ways. They can act as a pre-set and thereby directly affect whether stress is experienced as eustress or distress. They also can come to the fore as a direct reaction to the stress response itself and then, secondarily, influence the progress of the response. In either case, they act as modifiers of the stress reaction resulting all too often in such outcomes as anxiety, burnout, feelings of helplessness and hopelessness and a whole host of mental and physical illnesses.

The Positive Effects of Stress On Disease

In spite of the fact that Selye and others have recognized that there are positive aspects to stress (eustress), most people consider the term stress as implying something destructive. They view it as the cause of mental and physical problems. They do not often think of stress as inhibiting or counteracting disease. Yet, in many situations, it has just that capability. Selye (1976) pointed out in 1956 that though cancer may develop in body areas exposed to chronic irritation, "general stress tends to suppress cancerous growth." More recently he noted that "stress can inhibit or even prevent cancer" (Selye 1979). However, the stress that tends to reduce disease is stress which is usually of short duration or positive (eustress) in nature. Prolonged (chronic) stress or distress tends to have a negative effect on disease processes. The positive effect of stress on disease stems from the increase in body resistance in response to the initial alarm reaction. In many instances, this action is very selective in nature, only transferring its effect to combat very specific diseases.

Part of the selective positive influence on disease is due to the fact that, in general, an increase in body resistance to a particular stressor results in a decreased resistance to many other stressor agents (Selye 1976). It is also due to the combined dual response of the body to a given stressor, namely its specific and nonspecific effects. The nonspecific effect is similar regardless of the stressor. The specific effect, which differs from stressor to stressor, often tends to modify the nonspecific effect making its carry-over more localized in nature.

An example of this use of specific stressors to fight disease can be found in the old-time use of intramuscular injections of so- called foreign proteins. At the turn of the century, when this technique was most popular, milk was the most commonly used foreign protein. The procedure acted by creating a fever as the body fought off the foreign agent. The theory was that this reaction would then aid the body as a whole in combatting the resistant ailment in question.

Although the use of foreign proteins went the way of bleeding with leeches and cupping, we can begin to view these procedures as examples of how stressor agents have been used to activate the stress response to combat disease. Similarly, it is not unreasonable to consider the effects of liniments, mustard plasters and even vaccinations in the same light. Vaccinations serve the purpose of introducing a mild version of a given disease as a stressor in order to increase specific defenses of the body so that the system can cope better with a full-fledged version of the disease. This same principle may also be one of the factors behind the practice of homeopathy as developed by Samuel Hahnemann over a century ago.

There are two major tenets in homeopathy (Hahnemann 1961; Vithoulkas 1971; Coulter 1972,1973). The first ("like cures like") is that a specific disease is treated best by provoking the very symptoms of the disease. The drug that is used in therapy is designed to be as similar as possible to the disease itself so that it meets no resistance from the body, while at the same time gently stimulating the body's defenses to fight the disease.

The second tenet is that the smallest possible dosage is considered to be the most effective since it will not tend to aggravate the symptoms of the existing disease to any degree nor will it tend to create new symptoms.

It is interesting how a different perspective, in this case using the stress concept as a base, can offer new insights into total body operation and traditional therapies. The view that the adaptive response can be channeled through specific modifications of stressor agents may be one that can also explain some of the positive results achieved when using such techniques as Jacobson's progressive relaxation, confrontation therapy and even certain types of physical activities such as jogging.

The Negative Effects of Stress On Disease

Stress may be shown to have selective positive effects in inhibiting or delimiting disease, but more often than not stressful situations can increase one's vulnerability to illness and, under certain circumstances, can even cause sudden death. It can also contribute to an increase in one's vulnerability to everyday mishaps, accidents and errors in judgment.

It cannot be emphasized enough that stress itself is not necessarily a negative factor. A parachute jumper who enjoys the sport does not undergo the same stress reaction when jumping as someone who is compelled to jump for the first time because of a life-threatening situation. An individual who is pushed off a diving board certainly undergoes a different stress reaction than someone who practices diving as a skill. A major element in the stress response, therefore, is one's perception of the event and how one responds to the given situation.

A second major element is one's coping style or how one has learned to deal with the environment and stress. This learned response usually becomes so conditioned that the main reaction to stress often tends to take place during the anticipation phase rather than during the event or the period following it (Rutter 1981).

Although specific physical stressors such as an injury, infection, environmental toxins, extreme climate changes, etc., can trigger a stress reaction, more often the major factors involved in distress are psychological, psychosocial, emotional or attitudinal, namely, feelings of helplessness, hopelessness, loss and despair. It is these factors that appear to be related to so many common diseases (Grace & Graham 1952).

Respiratory infections, for example, have been found to occur more frequently after stressful life events (Jacobs et al. 1970). Gastrointestinal ulcers, one of the main stress problems demonstrated in Selye's Triad (see Figure 21), have been found to be more prevalent among those who have supervisory roles such as foremen than in those who are either craftsmen or executives (Dunn & Cobb 1962). They also occur almost twice as often among air traffic controllers who work at high-stress control centers than among those working at low-stress air control centers (Cobb & Rose 1973).

The immune system that is designed to offer resistance to all diseases in which the body must recognize "self" from "nonself," sparing the one and destroying the other, is also sensitive to stress (Borysenko 1984). In a research study using rats, it was found that a graded series of electric shocks produced a progressively greater suppression of lymphocyte functions that was in direct proportion to the intensity of the stressor (Keller et al. 1981). In a second study, rats were exposed to both escapable and inescapable shock. Lymphocyte production was suppressed in those rats exposed to inescapable shocks indicating that the psychological aspects of controllability rather than the stressor agent itself is critical in modulating immune system functioning--even in animals (Laudenslager et al. 1983) Jemmott and Locke (1984), reviewing the literature, concluded that even though more research is certainly needed, the bulk of evidénce also favors the view that psychosocial variables play a role in modulating the human immune response.

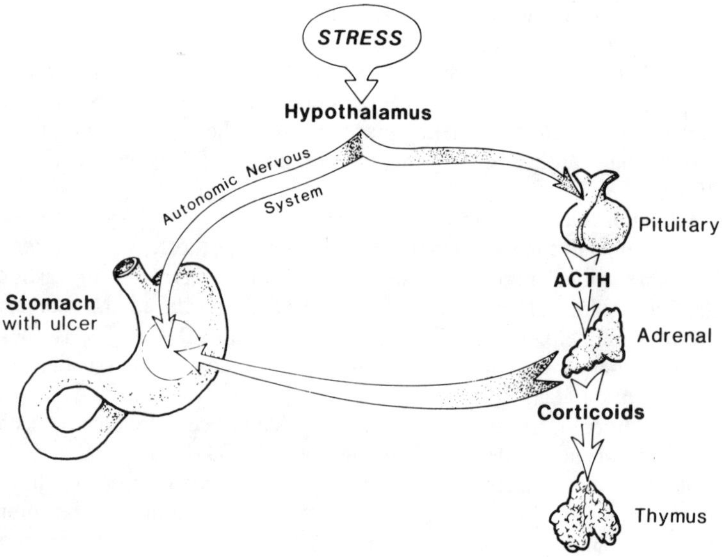

Fig. 21. Stress and ulcer formation

Cardiovascular Disease

It is estimated that heart disease takes the lives of millions of people every year including over 1 million Americans. Repeatedly, psychological stress has been implicated as being one of the major causes of the problem (Eliot 1979, Glass 1977). Psychological stress has been found to play a critical role in long-term blood pressure regulation by inducing changes in kidney function (Light 1983). Social stress has been found to influence the formation of atherosclerosis and atherosclerotic lesions in blood vessels even when serum cholesterol and other lipids were normal (Kaplan et al. 1983). In addition, many large-group studies have implicated stressful life experiences in the development of cardiovascular problems (Dohrenwend 1974, Jenkins 1976).

In recent years, due to the work of Friedman and Rosenman (1974), there has been general acceptance of the "Type A" behavior pattern as being a major contributor to coronary heart disease. Type A people are those who are unusually aggressive, impatient, competitive, hostile, insecure and work-oriented and who have a constant sense of urgency about what they do (Friedman & Rosenman 1959). There have been many studies that have shown that those with Type A characteristics have a much higher incidence of heart attacks than do "Type B" individuals who demonstrate fewer of these characteristics, being more slow-moving, less rigid and rarely hurried (Rosenman 1978).

To Friedman and Rosenman, a person's behavior pattern, either Type A or Type B, reflects the way he relates to life. In a way, the Type A/Type B classification has certain similarities to Selye's personality classification of "racehorses" and "turtles." There are also distinct differences.

To Selye, both "racehorses" and "turtles" are acceptable personality styles. The difference between them is in their operational tempo. An individual has difficulty only when he is in a situation (either hustle and bustle or quiet and slow- moving) that is not compatible or congruent with his preferred operational tempo.

To Friedman and Rosenman, Type A/Type B individuals reveal not only opposite preferred tempos but also completely different intensity styles in how they approach life. Though the combination of fast tempo and high intensity as reflected in Type A behavior is quite common today and is considered by some to be an advantageous ingredient for achievement in a competitive culture such as ours, in terms of its physiological effects, it appears to be the least desirable personality style to have.

It must be understood that the designations Type A and Type B represent the extremes of a behavioral continuum. Very few individuals have all the characteristics that represent either extreme or have them to the same degree. Yet, to understand where an individual might fit in this Type A/Type B continuum, it is important to be aware of the typical behaviors that those on the extreme ends tend to exhibit. Friedman and Rosenman

(1974) describe Type A and Type B persons as usually tending to demonstrate the characteristics listed in Table 6 and Table 7.

Table 6.

Typical Type A Behavior Patterns

1. A tendency to speak hurriedly and explosively.
2. A tendency to move and do things rapidly including working, walking and eating.
3. A tendency to be impatient such as with the slowness of events or with the actions and speech of others.
4. A tendency to do or think about two or more things simultaneously (talking to oneself while shaving, driving, during a conversation or when eating).
5. A tendency to maneuver conversations to the things one is interested in. If not successful, to look attentive while thinking about something else.
6. A tendency to feel guilty when relaxing or when there is nothing to do.
7. A tendency not to be observant.
8. A tendency to be overly preoccupied with things worth having rather than things worth being.
9. A tendency to schedule more and more in less and less time (a chronic sense of time urgency).
10. A tendency to react aggressively especially to other Type A persons.
11. A tendency to use characteristic gestures for emphasis such as talking with one's hands, clenching the fist or pounding on a table.
12. A tendency to use obscenities.
13. A tendency to having tics that reflect an inner struggle (rapid eye blinking, lip clicking, a jarring laugh).
14. A tendency to believe that success comes from getting things done faster than others.
15. A tendency to want to win and dominate at both work and play.

Table 7.

Typical Type B Behavior Patterns

1.	No sense of time urgency.
2.	No noticeable impatience.
3.	No harboring of free-floating hostility.
4.	No compulsion to display or discuss one's achievements unless it is appropriate to the situation.
5.	Can relax and have fun without any display of superiority.
6.	Can relax without guilt.
7.	Can work without agitation.
8.	In general, Type B persons feel that their achievements match or exceed their own expectations.
9.	Can more freely express affection.

Tables 6 and 7 reprinted, with permission, from M. Friedman and R. H. Rosenman, *Type A Behavior and Your Heart* (1974).

Friedman and Rosenman (1974) also determined that, as a group, Type A people demonstrate a chronic high-level sympathetic nervous system arousal response to stressors in which there is an increased secretion of adrenalin and cortisol. They found that extreme Type A people typically exhibit an increased blood level of serum cholesterol and fats, a marked lag in eliminating ingested (exogenous) cholesterol, diabetic or prediabetic traits, a narrowing of the small capillaries, an increased tendency for blood-clotting elements (platelets and fibrinogen) to precipitate out, and a tendency to have higher blood pressure than Type B individuals. They also found that Type As tend to exercise less because they cannot find the time. When they do exercise, they do it with a vengeance, as if they were driven. They also tend to smoke more cigarettes and eat more foods rich in cholesterol and animal fats.

Redford Williams (1982) and his associates noted that during mental arithmetic tasks, Type A subjects showed greater muscle vasodilation and produced increased levels of norepinephrine, epinephrine and cortisol than Type B subjects. On reaction time tasks, Type A subjects secreted excessive levels of testosterone. Both cortisol and testosterone are known to be related to increased hardening of the arteries.

Holmes et al. (1984) found a more limited form of sympathetic nervous system arousal in Type As. They noted that the difference in physiological arousal between Type A and Type B behavior patterns came to the fore only when subjects were in a highly challenging situation such as when involved in an extremely difficult task. They also found that the difference in physiological arousal between Type A and Type B subjects was limited only to systolic and not diastolic blood pressure, pulse rate, pulse volume or skin resistance. Their research therefore indicates that Type A behavior is related only to specific aspects of what is considered to be sympathetic arousal and not to physiological arousal in general.

Though there are some Type As who appear to be less at risk than some Type B individuals (Eliot 1982) Friedman and Rosenman's Type A behavior pattern is now quite established as a major risk factor for coronary heart disease. It achieved its current prominence and level of acceptance mainly because the traditional risk factors for heart disease such as hypertension, high serum cholesterol and cigarette smoking were only capable of predicting about half of the new cases of clinical heart disease (Jenkins 1971,1976).

Williams (1984) attempted to isolate the most significant or "toxic" element among those factors that compose the typical Type A behavior pattern. These include speed-impatience, ambition- competitiveness and hostility-anger. He determined that the key factor that related most to the onset of heart problems was the hostility component. In fact, hostility was a valid indicator of heart problems even when the Type A profile was not.

Williams then tried to determine what hostility actually represented. To do this, he analyzed the hostility (Ho) scale of the Minnesota Multiphasic Personality Inventory (MMPI) and concluded that "cynicism" rather than "hostility" was the more appropriate name for what the Ho scale measured and that this term "better, perhaps, than any other single word, captures the toxic element in the Type A personality." Williams also noted that people who do not trust others tend to be more vigilant and that this cynicism can become a self-fulfilling prophecy since "a hostile, untrusting person often elicits hostile behavior from others, thus confirming his suspicions about them." This only makes him more antagonistic, further increasing adrenalin and cortisol secretion, which, if it becomes frequent and intense, can then lead to coronary disease.

Over the years, many methods have been used to determine who is a Type A person (Glass 1977, Matthews 1982). The original method of classifying Type A/Type B behavior was through the use of a structured interview developed and standardized by Friedman and Rosenman (Rosenman et al. 1975). In this interview, the subject is asked approximately twenty-five questions dealing with his ambitions, competitiveness, sense of time urgency and the nature of any hostile feelings he may have. The interviewer deliberately phrases the questions to create a stressful atmosphere in order to reveal the characteristic way the subject reacts, at least in terms of his speech patterns.

A second method involves the use of an objective self- administered questionnaire developed by C. David Jenkins (1974) called the Jenkins Activity Survey for Health Prediction (JAS). Jenkins also introduced a modified version of the questionnaire specifically designed for use with college students.

A third method of determining Type A behavior is through the use of the ten-item Framingham Type A Scale which assesses an individual's competitiveness, sense of time urgency and his perception of job pressures. The test questions were originally selected from a 300-item inventory used to assess the psychosocial attributes of subjects enrolled in the Framingham Heart Study (Haynes, Feinleib & Kannel 1980).

All three of these measures appear to be reliable, but Matthews (1982) found that the association among them "is less than impressive" and that the three measures "have only the slightest margin of overlap. Matthews, therefore, concludes that these three measures of Type A behavior "assess different behavioral characteristics," yet "nonetheless, all are related to the incidence of coronary heart disease and assess some of the behaviors collectively called the Type A behavior pattern."

Friedman (Friedman & Ulmer 1984) has expressed the view that the Jenkins Activity Survey is of "dubious value" for three reasons. First, it is based only on a preliminary and less-developed version of the original Friedman and Rosenman oral interview. Second,it depends on the subject's own appraisal of himself rather than on the more revealing observations of overt Type A behavior as would be exhibited by facial, vocal and bodily manifestations. Finally, subjects are often notoriously poor observers of their own actions and motives and for that reason the questionnaire is not as objective as many who use it regard it as being.

Glass (1977) suggests that the collective characteristic behaviors of Type A people mainly represent the attempt on the part of these individuals to control events that they perceive as being uncontrollable and potentially harmful. His view considers Type A behavior as being methods of coping with perceived stressful events. Type A individuals tend to try much harder in situations where others will either coast or give up. Because they try so hard, if their efforts meet with repeated failure, they tend to give up and act helpless, feeling that nothing will really help. It is this feeling of helplessness that contributes to both depression and to coronary heart disease.

Friedman and Ulmer (1984) feel that the two basic components that trigger Type A behavior are a covert insecurity or inadequate self-esteem and hyperaggressiveness. It is usually one or both that causes the struggle to begin. They consider that the most likely origin is the feeling of insecurity that comes from the failure of the Type A person in infancy or early childhood "to receive unconditional love, affection and encouragement from one or both of his parents." This means love simply for being himself and not as a reward for some specific achievement. Male Type As are more sensitive to a shortage of maternal love and emotional support while female Type As appear more sensitive to the paternal relationship.

As the Type A struggle continues, it fosters the emergence of time urgency or hurry sickness and free-floating hostility. If this struggle persists and is severe enough, it leads to a tendency toward self-destruction manifested by a deterioration of the personality, emotional exhaustion and pathophysiological disease processes such as clinical coronary heart disease.

Hot Reactors

Rather than determining Type A behavior during the course of a structured interview or as a result of a questionnaire, Robert Eliot and his associates focused instead on purely physiological rather than behavioral responses. They used a variety of specific challenging tasks such as striving to achieve on an electronic video game, a mental arithmetic task where one has to subtract seven from 777 serially as fast as possible for three minutes, and the use of a cold pressor test where one's hand is placed in ice water.

Those who have a normal physiological cardiovascular response to these stressful situations are considered to be "cool reactors." Those with an abnormal cardiovascular and general stress response are considered to be "hot reactors" (Eliot & Buell 1983, Elliot & Breo 1984, Brody 1983). Eliot found that one can be Type A according to behavioral tests without exhibiting physiological changes in blood pressure, cardiac output and blood vessel resistance under stress. In other words, one can exhibit Type A behavior and still be a "cool reactor." A "hot reactor" and Type A behavior do not necessarily go hand in hand. It is possible to be psychologically intense (Type A) but not physiologically intense (a "hot reactor").

Outward behavior, therefore, does not always reveal how one responds to stress physiologically. This can serve to explain why some Type A individuals do not get heart disease while some Type B individuals do.

Learned Reactions

Matthews (1982) feels that some children start out with a high energy level, are more active physically and more responsive emotionally (Brody 1984a). When these children are placed in a social setting where parents and teachers encourage achievement and productivity they are more likely to begin to demonstrate Type A behaviors early in life. Friedman and Ulmer (1984), as mentioned before, feel that Type A behavior stems from the child's perception that his parents' love depends mainly on his achievement. Eliot feels that being a "hot reactor" is similarly, carefully learned.

All agree that what is learned can be unlearned or at least modified (Brody 1984c, Stroebel 1982). As Stroebel (1982) puts it, it should be possible to maintain the high-performance traits of the Type A personality without devastating one's physical health: to have a Type A mind in a Type B body. If this can be done, and many agree that it can, there is an excellent chance that heart disease can be prevented, minimized or delayed (Brody 1984b). If changing behavior can have such a positive effect on those with overt heart disease, it should be even more helpful preventively for those who exhibit no apparent physical problems.

Chapter 13

Stress, Attitude And Disease--II

The vast majority of scientifically trained individuals are so deeply grounded in a concrete approach to physiological function that they often find the concept of mental states to be too intangible to be considered as a major cause of disease. They find it less disturbing to think of mental states as concurrent conditions that can aggravate a particular disorder or as secondary complications arising from the disease itself.

Over the years, however, there has been a reluctant but growing recognition of the role of the mind, attitudes and emotions in disease. More people are accepting the role of the emotions in the formation of ulcers. There has also been an increased acceptance of behavioral factors such as the Type A/Type B personality pattern classification in relation to heart problems. Yet, even in this instance, Friedman and Ulmer (1984) point out that in a recent 1,943-page text on heart disease, only four paragraphs discuss Type A behavior as a possible risk factor. More discerning individuals today are also questioning the enigma of the placebo effect where a "sugar pill" often mimics the positive action of a specific drug simply because the patient apparently believes that it will help. What better example than this can there be to illustrate the importance of attitudes and beliefs on disease processes?

Cancer and Stress

Many who accept psychological or mental factors as being possible causes of ulcers and coronary heart disease still tend to think in more limited terms when it comes to cancer. They feel more comfortable with the view that malignancy results solely from chronic irritants, toxins, viruses, radiation, oncogenes, age or genetic factors. Even Ader (1982), who stresses the relationship between the mind and immunity, becomes very cautious in regard to cancer, feeling that there are just too many unknowns in the equation (Anderson 1982).

One part of the problem is that modern scientific and medical beliefs are still strongly influenced by Descartes' philosophy of dualism. This philosophy considers mind and body to be completely separate aspects of the organism. A further extrapolation is that bodily disease must be unrelated to mental states.

In spite of these influences, there has been a slowly developing body of evidence that points to a possible relationship between cancer and psychosocial stress. There have been a number of animal studies that highlight this relationship. While some of the results appear contradictory, there are others that appear reasonably consistent and significant (Sklar & Anisman 1981).

Sklar and Anisman (1979) report a study in which laboratory mice were injected with living tumor cells. After twenty-four hours, the mice were placed in a shock apparatus and given a six-second shock once a minute for one hour. The animals treated this way revealed a significant decrease in the time required for tumor development, an increase in tumor size, and a decrease in survival time as compared to a control group which did not receive any shocks.

A second experiment involved two groups of mice. One group was allowed to escape the shocks through a hatch that opened five seconds after the shock started. Those in the second group could not escape. The result was twofold. First, it was noted that stress influenced tumor development in general. Second, it was revealed that the effect was dependent on whether control over stress was possible. Inescapable stress was seen to accelerate tumor growth. This indicated to Sklar and Anisman that the inability to cope with stress behaviorally rather than physical stress, per se, is ultimately responsible for the effect that shock has on tumor size.

Human studies may be somewhat less exacting but equally revealing. As far back as the second century, the Greek physician Galen observed that depressed, melancholy women were more prone to develop breast cancer than were those who exhibited a cheerier and more sanguine disposition. One problem in establishing these kinds of relationships in humans is that it is not considered ethical to inoculate human subjects with cancer cells or intentionally expose them to active carcinogens for the purpose of research. Therefore, almost all studies of human cancer tend to be retrospective, dealing with events that have already occurred though the condition may have taken years to develop. This makes it difficult to isolate all possible causative factors.

One important prospective study of emotional factors that precede disease was undertaken by Carolyn Bedell Thomas (1973) of the Johns Hopkins University School of Medicine in 1946 (Thomas, Duszynski & Shaffer 1979; McQuerter 1978; Holden 1978).

Thomas and her associates psychologically surveyed 1337 students graduating between 1948 and 1964, assessing their family relationships, behavioral characteristics and responses to Rorschach and figure-drawing tests. She then had them complete yearly questionnaires concerning their health and living habits.

Thomas was able to correlate given personality traits with major diseases noting that "adverse family attitudes were associated with later diseases, especially major forms of cancer." She also noted that her data indicated that cancer tends to strike those persons who are low-key and nonaggressive and who keep their emotions to themselves. She observed that most cancer victims were lonely persons who had not been close to their parents. A lack of closeness to the father during childhood appeared to be a major risk factor for lung cancer.

Holden (1978) reports that Nicholas Rogentine, an immunologist at the National Cancer Institute, found that patients successfully operated on for malignant melanoma who relapsed tended to minimize the significance of their illness. This led to the conclusion that repression and denial is apparently related to a poor prognosis.

D. M. Kissen and associates (1963,1969,1967,1962) performed psychological studies on over 1000 men with undiagnosed pulmonary disease, all of them smokers. Those found to have lung cancer appeared to display an inability to express emotions. In general, they seemed to suffer from emotional denial and repression as compared to those found to have other types of lung disease. Kissen (1964) also noted that the more repressed the individual, the fewer cigarettes it took to induce cancer.

Schmale and Iker (1966) were able to achieve statistically significant predictions (36 cases out of 51) as to which women with suspicious cell changes in the cervix would develop cancer. This was based solely on the degree of hopelessness and helplessness that the patient revealed in interviews and on the Minnesota Multiphasic Personality Inventory (MMPI).

Abeloff and Derogatis (1977) found that women with metastatic breast cancer who could express their hostility survived longer than those who were compliant and unassertive. Similarly, Greer et al. (1975) found that women with breast lumps who suppressed their hostility revealed a higher percentage of breast cancer. Those who exhibited a "fighting spirit" or who were absolute "deniers" appeared to have a statistically significant advantage over those who were stoic acceptors and who basically felt helpless and hopeless (Greer & Pettingale 1979, Greer 1979). Bahnson and Bahnson (1964,1969) also found strong indications that a relationship existed between repression and denial and malignancies.

One of the major modern pioneers relating cancer with personality and emotions is Lawrence LeShan (1977,1959,1966; Le-

Shan & Worthington 1956; LeShan & Reznikoff 1960), whose scientific endeavors in this area started over thirty years ago. From 1952 to 1964, LeShan tested more than 400 cancer patients with the Rorschach (ink blot) test, the Thematic Apperception test and the Worthington Personality History. The latter allowed patients to tell their life stories in their own words, thus revealing their ego defenses and their everyday techniques for coping with life. LeShan also did more than 100 hours of intensive psychotherapy with each of 71 terminal cancer patients. He found that 72 percent of the patients (as compared to 10 percent in a control group undertaking psychotherapy for other reasons) suffered the loss of a central relationship within a period of eight years to a few months of the onset of the disease. Most lived for years with feelings of isolation, despair, desertion and loneliness and found it difficult to express anger or hostility. This pattern suggested to LeShan that the tumors may be the somatic offshoot of an emotional death wish reflected right down to the cellular level (McQuerter 1978, Holden 1978). In a similar vein, Booth (1969) feels that cancer patients, as a group, are more pessimistic, despairing, and more vulnerable to object loss and that in their depressive reaction to loss they literally turn part of their bodies into the lost object.

Disease as a Mind-Body Process

Too often there is a tendency to think of a disease as a particular entity and not as part of a total life process. In this way, consciously or unconsciously, the person is separated from the illness. Even though sickness is concerned with organs, tissues and body parts, "health is a domain in the realm of consciousness and of meaningful behavior...it has to do with the feeling of life and the fullness and color of it" (LeShan 1982). Surprisingly, some of the most dedicated disease-oriented scientists have had a broader outlook than most people realize. Louis Pasteur dedicated a major portion of his life to uncovering specific germs or bacteria as causative factors of disease. Yet even he was aware that not everyone exposed to a given germ would develop the particular disease that it was supposed to carry. He accepted that other factors including one's emotional state determined whether one fell victim to a disease or not (Dubos & Esconde 1979).

Claude Bernard, the famous physiologist and philosopher, wrote that "Illnesses hover constantly about us, the seeds blown by the wind, but they do not set in the terrain unless the terrain is ready to receive them." Before his death, Louis Pasteur's last words emphasized his agreement with this concept, "Bernard is right," he said, "The germ is nothing. The terrain is all" (Dubos 1979).

Barbara Brown (1980), author of Stress and the Art of Biofeedback, has stated that disease originates in intellectual processes and that stress has its effects on health because the rational mind confronts situations with which it cannot cope.

Elmer Green et al. (1970) state that "every change in a physiological state is accompanied by an appropriate change in the mental-emotional state, conscious or unconscious; and conversely, every change in the mental-emotional state, conscious or unconscious, is accompanied by an appropriate change in the physiological state."

Another ramification of the mind-body relationship is the tendency to "brace" oneself during what is perceived to be a potentially painful physical or psychological situation such as getting a tooth drilled or being inoculated. The neuro-muscular- hormonal changes accompanying this reaction only tend to increase whatever physical and psychological discomfort would normally result from the situation. Whatmore and Kohli (1979) call this muscular bracing dysponesis. Dysponesis becomes a "trap" where tightening up one's body or body parts, to put up a brave front to the world in order to appear more mature, results in an aggravation of the original problem. In addition, the body part that is more persistently braced often tends to become the source of specific "diseases" such as lower back disorders, temporo- mandibular joint malfunction and migraine headaches. It may also be possible that selective dysponesis plays a role in the specificity of certain types of cancer.

Eliot and Breo (1984) emphasize that disease stems from "the feeling that life is a joyless struggle, the lingering sense of loss and anger turned inward." They consider that these attitudes come from the conflict between what one really wants or realistically can achieve and the tendency to conform to what society, not the individual, values. This creates an uncertainty which results in life appearing empty and the future hollow. "These people are anxious and angry about losing control of their lives...(and)...often avoid turning their anger outward, for fear of losing even more. Instead, they turn it inward on the only 'safe' target--themselves."

To Eliot and Breo, stress is only a burden if we allow it to be through the belief that we are helpless in the face of life's forces. But they emphasize that this is only a belief and not a fact.

They also differentiate between an "acute alarm" stress reaction by which the body prepares for "fight or flight" and a "chronic vigilance" reaction during which the body prepares for long-term endurance. They note that the alarm and vigilance reactions are triggered by different perceptions of events. Alarm is seen to occur when one perceives a challenge to control. Vigilance is seen to occur when one feels a loss of control possibly because of self-doubt, a sense of failure, or a feeling of being entrapped. Chronic vigilance and the attitude and beliefs it reflects is seen as being a major cause of disease.

Cancer and Imagery

In spite of prevailing opinions to the contrary, there is a growing body of literature that appears to support the relataionship among the factors of stress, attitude and disease with attitude apparently being the most important factor.

If this relationship is valid, then, as difficult as it often may be to accomplish, inducing a positive change in attitude should have a corresponding positive effect in reducing distress and reversing existing disease processes.

The clinical psychological, psychiatric and psychoanalytic literature of the past half-century has been filled with individual case histories and group studies where counseling and psychotherapy have been shown to eliminate or aid many specific diseases. Skeptics, however, insist that much of this is anecdotal with the results probably being coincidental or due to variables that have not been properly considered. Some consider that physical conditions that improve through psychological therapy or changes in attitude are not true diseases but functional problems that simply mimic true diseases. Many of these doubters have difficulty assessing some of the findings in the field of psychoneuroimmunology or even the positive results on coronary heart status by reversing Type A behaviors.

It would seem that a more substantial and critical disease entity such as cancer, that can be medically analyzed and verified, would offer more convincing evidence of a possible relationship to attitude. Yet, here too, not everyone is convinced by the reported studies. Although many consider it to be quasi-medical, there has been an increasing interest not only in the relationship of attitude to cancer but in the course of attitudinal therapy on the control and conquest of malignancy.

Outside of stress reduction techniques, relaxation procedures and counseling, one of the more recent and novel approaches to cancer treatment involves the use of visual imagery. It is an approach that appears to be quite effective in returning the patients' control or feeling of control over their illness and, by extension, over their lives (Forrest 1981c, Hall 1984, Pelletier 1977).

The use of visual imagery or visualization in cancer therapy started with the work of O. Carl Simonton. Simonton was originally in charge of radiation therapy at Travis Air Force Base in California where he noted a higher incidence of remission in those cancer patients who appeared to be more excited and enthused about their treatment program and who seemed to believe deeply that the therapy would work.

Simonton and his wife, Stephanie Matthews-Simonton (1975), decided to utilize positive attitude training in an attempt to remove the precipitating emotional cause of the disease and thereby hopefully inhibit its progress. Their initial work involved the modification of the belief systems of those patients and their families who considered their condition hopeless. At first, they experimented with biofeedback and autogenic therapy. Later, they stressed relaxation therapy and imagery exercises in which the patient actively visualized the destruction and elimination of the diseased tissues.

In their current approach, the patients are first taught a modified form of autogenic therapy with an emphasis on breathing (Simonton & Simonton 1978). They are told to repeat the word "relax" silently and to just "let go." When they feel totally relaxed, they are instructed to image their illness, visualize the specific therapy that is being used and how they imagine it is affecting the tumor, visualize the action of their own autoimmune system as it comes to their defense and finally, to image the tumor shrinking as the white blood cells, chemicals, radiation, etc., successfully attack the invader without harming healthy tissue.

The Simontons feel that the imagery/relaxation component of their program works by enhancing the action of the patient's own immune system in combatting the cancer. They have identified eight important features in their approach, noting that patients who did not do well in their program often lacked one or more of the following critical features:

1. The cancer cells should be imagined as being weak, confused, soft and easily broken down. Patients are encouraged to imagine cancer cells as pieces of hamburger meat or fish eggs rather than more tenacious objects like ants or crabs.
2. The treatment should be positively viewed by the patient as being "strong and powerful" and fully capable of destroying the cancer.
3. The healthy, noncancerous cells should be imaged as being able to repair easily any damage done by the therapy.
4. The body's immune system (the white blood cells) should be imagined as greatly outnumbering and overwhelming the cancer cells.

5. The white blood cells should be viewed as being aggressive and quite eager to destroy the cancer cells so that their ultimate victory over the malignancy appears inevitable. Sometimes it is helpful to visualize the white blood cells as vicious dogs, rats or soldiers.

6. The dead cancer cells should be visualized as being flushed out of the body in a biologically natural way.

7. Patients should consistently visualize themselves as being totally healthy and disease-free.

8. Patients should visualize themselves freely fulfilling their life's goals.

Achterberg and Lawlis (1978) have successfully used a similar imagery approach for cancer therapy. They also developed a test instrument, called "Image-Ca," to evaluate the effectiveness of a patient's imagery and also to predict the future development of the disease. The test measures 14 dimensions of imagery with each dimension rated on a five-point scale. Some of the factors that are analyzed include the imaged strength of the cancer cells,the immune system and the medical treatment as well as whether the images are concrete or symbolic in nature. The scoring also includes an estimation of image vividness, image activity and the degree to which the imagery is practiced.

The testing is done by first having the patients draw the images they conjure up of their tumors, white blood cells and medical treatment. They then go through a structured interview so that the 14 dimensions can be scored on the five-point scale.

Achterberg and Lawlis determined that the imagery of the white blood cells in particular (vividness, activity, size and strength) appear to be more important in determining a short-term prognosis of the disease than the other aspects of the "Image-Ca" test. They also feel that while blood chemistries might reflect the body's current status, psychological factors such as the "Image-Ca" test seem capable of predicting future physical response patterns (Holden 1978).

Philip Lansky found that some patients have too much unresolved feelings of anger, hate, grief or loss to get maximum benefit from using aggressor-type images in fighting cancer cells. Lansky is reported as noting that "when the disease reflects undischarged emotional energy turned back upon itself, combative visualizations may reinforce such denial" ("Cancer: Positive Images" 1982). He proposed instead that repressive patients should "love" their cancer rather than attempting to kill it, recommending "a visualization that accepts and 'melts' one's cancer via a healing gold light emanating from the bone marrow...(which) is the physical source of immunologic competence."

Finally, returning to those who are more skeptical, Constance Holden (1978) quotes Jimmie Holland, Chief of Sloan-Kettering's psychiatry service, as saying that things have gotten out of hand, that "there are no scientific data linking psychological factors to cancer, either in predisposing a person to the disease or in determining how well a cancer patient will respond to treatment." Holden also quotes Holland as believing that the Simontons and their imagery approach are perpetrating a "cruel hoax."

Correlates of Attitude and Disease

Of the great number of possible diseases, so many are highly selective, affecting only certain organs or body parts. Even - malignancy is localized in nature, affecting specific tissue structures such as breasts, cervix, skin, lungs, spine, testicles, ribs, blood, lymphatics and brain. This selectivity has resulted in some of the advocates of the role of attitude in stress and disease trying to reconcile specific attitudes with specific disorders.

Some have developed or assembled psychological profiles that seem to be found with different diseases. Arehart-Treichel (1980) suggests that the following relationships exist:

Those with ulcerative colitis tend to be emotional, sensitive, bright, and overly dependent on parents. Those who suffer from ulcers tend to be hardworking, meticulous, ambitious and somewhat self-centered, internalizing rather than confronting issues and fantasizing their bodies as being weak and vulnerable. They also exhibit a need for affection. Those with migraine are seen to be duty-driven and guilt-ridden while tending to repress negative emotions. Those suffering from eczema are seen to be exceptionally sensitive, tending to repress their emotions. They also appear to have been stroked a lot by their parents as well as having slept with them. There are additional profiles for those with hayfever, asthma, hives, angina, hypertension, heart attacks and cancer. Unfortunately, there is a great deal of redundancy in the profiles and many of the aspects listed appear overly general. This would indicate that even if the premise is valid, much more research is needed to refine the relationships.

Others have assumed that a given disease entity is directly related to at least one basic core attitude that usually can be expressed verbally (King 1981). This view infers that many diseases are often metaphors for possible verbal statements about an individual's attitude to life or to persistent life events (see Table 8).

These relationships are also broad and general. Yet, again, if the basic premise has any validity at all, it indicates a promising direction for future, more sophisticated research.

Table 8.

Verbal Metaphors for Disease

Statement	Disease
"Life gives me a headache."	Headaches, migraine
"Life is choking or stifling."	Difficulty breathing, allergy, lung problems
"Life is a pain in the neck."	Stiff neck
"I can't stomach that; I can't digest that; Life just eats me up."	Gastrointestinal problems, ulcers, etc.
"I can't swallow that."	Throat problems, difficulty swallowing
"I can't bite into things."	Dental problems
"I can't take what I hear."	Hearing difficulties
"I know I'm spineless."	Spinal problems
"I'm always bending over backwards."	Back problems

Reprinted from Jane E. Brody, "Heart Attacks and Behavior; Early Signs are Found," *New York Times,* Aug. 7, 1984.

Pain and Attention

Most people at some point or other have been curious about the role that pain plays in human life. Philosophers, theologians, physicians, as well as the average individual, have been intrigued with this. For example, does pain serve a purpose or is it an anomaly, a quirk of nature?

In an earlier chapter, we differentiated between loneliness and being alone and between feeling crowded and being in a crowd. Similarly, we have to differentiate between the sensation of pain and the subjective experience of pain (Bresler & Turbo 1979). The former is the simple recognition that some form of unpleasantness exists. The latter is that same sensation modified by one's attitudes and beliefs.

Many believe that pain exists as a way of drawing conscious awareness to a particular disease, irritation or trauma, perhaps as a stimulus to do something about it. In this view, pain is a message for some appropriate action to be taken. Too often, we tend to think that appropriate action must be physical, such as removing oneself from the source of pain, taking medication to heal the cause of the pain or taking drugs to suppress it. However, since attitudes and beliefs are involved in the experience of pain, especially the chronic variety, pain may be a signal to change mentally the way one views life and life events.

It is often noted now, in so many instances, the body heals itself without help. Many diseases simply run their course and resolve themselves regardless of what intervention takes place. The old saying is that the common cold lasts seven to ten days whether or not medication is taken. In most instances, all that seems to be needed for healing to take place is a "tincture of time" and some patience.

If in most instances, the body tends to heal itself, why should there be a need for a built-in physiological drive such as that of pain to draw attention to an injury or disease? What is the message that pain is really geared to convey?

One view comes from Meldman (1970) who relates pain to self-directed attention. Meldman considers that there is a "biological and survival utility" in self-directed attention. He notes that attention can either help or hinder the body's repair mechanisms by directly influencing the adaptive response.

This concept is not out of line. It can be observed in many ways every day. Thinking about an itch, a toothache, a headache or the pain from a bruise often increases the symptoms to the point of making them unbearable.

On the other hand, there are those, such as Victor Frankl (1960), who feel that increasing one's attention to a functional problem often helps to minimize it. Frankl calls this "paradoxical intention." Concentrating on making oneself blush more, for example, helps to reduce one's blushing. Similarly, thinking about the size, shape and "color" of a functional headache often tends to reduce the dimensions and the severity of the headache.

Why is there this difference in effects? Why is it that attention can both increase and decrease the intensity of a headache?

The answer is in how the attention is applied. Specifically, the key is whether attention is related to curiosity and acceptance or is associated with the act of coping.

Coping with a problem, is, in effect, resisting or fighting it. Those who have a fear of enclosed spaces usually cope by avoiding elevators. When they do have to enter an elevator, they increase their fearfulness by internally verbalizing how bad they are probably going to feel, whether or not they are going to survive it, and why this has to happen to them in the first place. Actively avoiding tension and fear sustains the tendency to be tense and fearful. Actively fighting tension and fear tends not only to sustain but to strengthen the tendency to be tense and fearful. In other words, coping with a condition tends to reinforce and augment the condition.

Facing a problem, including pain, passively and without verbal judgment lessens the stress response. Simply observing the situation, such as the headache, without tension and even with some enjoyment (if it is at all possible) tends to have a more positive effect on how the individual adapts to the situation-- lessening the symptoms.

Kabat-Zinn (1984,1982) calls this the mindfulness approach in which one consciously directs one's attention to the sensation of pain while observing it dispassionately as just being an event in the field of general awareness. If one can separate one's feelings and thoughts from the sensation in question, then it is easier to accept the sensory component for what it is and in that acceptance, most people find that they suffer less. This is something that everyone can try for themselves.

The relationship between pain and attention is another example of the importance of the psychological state (one's attitude and behavioral pre-set) on the stress response.

Bresler (1984) offers another equally interesting perspective on the experience of pain and the message it conveys. He notes that "people who suffer pain, especially the chronic variety, typically have difficulty dealing with the process of change." Therefore, the message that persistent pain attempts to convey reveals the relationship an individual has to changing life events.

Bresler does not see pain as the cause of discomfort and suffering. Instead, when pain is persistent and is unbearable, it is the result of an inability to let go of the known, the old and the familiar and to be able to readjust to the changes thrust on the individual by life. To Bresler, the deciding factor in the way pain is experienced is in how one views his life and his world. As John Lilly (1972) so aptly stated, "In the province of the mind, what you believe to be true is true or becomes true...." In this way, even though starting from a different point of view, Bresler, similarly to Meldman and many others, also sees the perception or experience of pain as relating to one's attitudes, beliefs and expectations.

Chapter 14

Experience and the Role of Belief Systems

What has to be pursued is why so many people resist life as it is. Why do so many exhibit signs of helplessness and hopelessness? Why do so many feel that they cannot control events in their lives? Why are so many fearful? Why are so many anxious? Why do so many feel vulnerable? Why do so many have a poor self- image and low self-esteem? Why are signs of burnout so prevalent?

There is obviously more than one answer. At least part of the answer lies in two psycho-behavioral factors, both of which are fundamental to the stress response. The first is the role and status of experience. The second is the childhood origin of our pervading belief systems: the way one has been carefully taught by parents and society to perceive and to cope with events.

The Role of Experience

The use of the term "experience" has become commonplace today. It is usually taken to mean some form of participation, observation, sensing, perceiving or feeling. Actually, there often appears to be less interest in experience, per se, than in the ends for which experience is used--its goal, so to speak.

Piaget (1955,1952), for example, was primarily concerned with experience as the medium through which intellectual development takes place.1,2 His emphasis was on the development and action of thought processes. Experience, to Piaget, was simply giving the child the opportunity to deal with more aspects of the environment, thus allowing his developing mental structures to become more complex and more able to cope with environmental demands.

Jerome Bruner (Bruner, Olver & Greenfield 1966; Bruner 1966) views experience in terms of cognition. He differs from Piaget in seeing developing thought processes as being partly produced by environmental interaction (experience) and partly adaptive to it.

Piaget talks about the early sensorimotor stage and the later stage of concrete operations during which symbolic cognitive activity begins. Bruner considers that human beings go through an enactive stage where reality is represented by motor acts, an iconic stage where knowledge is represented by images, and a symbolic stage in which knowledge becomes represented by symbols which can represent both concrete information as well as abstract ideas. The terms used by Piaget describe how experience is used in mental organization. The similar terms used by Bruner describe how experience is represented at a cognitive level. Neither describes experience itself. It is the elements of experience, what we respond to when we experience, that we will be interested in. In these terms, we can consider that there are three types of experience: primary (reality-oriented) experience, secondary (symbolic) experience and tertiary (interpretive) experience.

Primary experience is a direct form of sensorimotor experiencing. When the primitive caveman Og reacted to a charging sabre-toothed tiger, he directly experienced and responded to an actual on- going event. He could see the tiger, smell the tiger, hear the tiger and even feel the change in air pressure as the tiger lurched at him. He was there! Similarly, if one touches a hot surface, steps on a nail, gets attacked, climbs steps, hears a car start, feels the wind against his skin or experiences a toothache, he is also responding to an actual on-going event. In fact, he is participating in the event. It is a true encounter. Primary experience, therefore, involves a direct one-on-one sensorimotor interaction to or with an event that an individual is directly involved in. What is seen, heard, smelled, tasted or felt is really there at the moment. It is pure reality-based sensation.

Secondary experience pertains to symbolic or substitute experiencing. Symbolic experience relates to what is not actually experienced at the moment in total sensorimotor terms. It is the use of symbols, both images and words, as a substitute for experience. For example, language (both written and oral) and pictures (in books, newspapers, magazines, photographs, movies and television) are symbols for experience. Symbols only represent experience. They are never the actual experience.

Though there are obvious advantages in using symbols, especially when dealing with abstractions, there are also some less obvious but equally, if not more important, disadvantages.

If we see pictures of starving children or mutilated bodies while we are comfortably seated at home, in school, in the movies, in the library or at work, we still have the tendency to react emotionally and physiologically as if we were actually there in space and time witnessing the real event. The same happens when we watch "live" coverage on television. The same reaction occurs when we read about events while not being in the actual situation. We may get angry when we read about an elderly person getting attacked. We may get fearful when we read of robberies in our neighborhood. Our hearts may beat faster when we come to a crucial part of an adventure story. In all these instances, however, the physiological reaction that occurs can be considered inappropriate since we are not ourselves participating in the event at that moment in time in true sensorimotor terms. It is not an actual event for us. There are also indications that the body somehow physiologically knows the difference.

Tertiary experience pertains to interpretive experiencing. It is not a pure form of experience as are reality-oriented experience, sensorimotor experience and symbolic experience but it is distinctive enough to deserve its own category. It is experience that is masked and distorted by a veneer of internal verbalization in which judgments and comparisons are constantly being made. Interpretations and judgments tend to color on-going events causing us to respond not to what is actually happening but to our perception or interpretation of the event.

The body is a fascinating instrument. It is designed to respond to actual events, to real occurrences. It adjusts automatically to changes in temperature, humidity, pH, blood-sugar levels as well as to sudden threats to the body. The lids blink and close if something approaches the eyes too quickly. We lurch backward if we daydream while crossing a street and almost get struck by a car. In most instances, the body responds appropriately to the particular situation and usually in reasonable time. This is biological intelligence at work.

The body also responds biologically to substitute events as in the use of symbols in, for example, reading or watching television and to nonevents such as one's assessment or perception of an event as well as to the overlay of fantasy and imagination onto reality. The problem is that the response in these instances is not in a one-to-one relationship with the event at hand. It is this dissonance between fact and fantasy that makes it stressful. A secondary problem that stems from this form of dissonance is that the individual who deals extensively with symbols or internal speech, situations that our society seems to cultivate, tends to lose the ability to differentiate effectively between the actual and the imagined. Eventually there is developed a tendency not to trust what is actually experienced since one begins to rely more on symbolic and interpretive sources of experience. Once one cannot trust what is actually sensed, a further tendency develops to be constantly on guard, alerted or aroused even to situations that are totally innocuous.

One also increasingly tends to lose the ability to respond as a participant to on-going events, becoming a more judgmental observer instead with a progressive inability to differentiate between what "is" and what "is not" while tending to react more to what "is not" than to what "is."

The Role of Belief Systems

The second psycho-behavioral factor fundamental to the stress response is the role played by our attitudes and beliefs. It is these attitudes and beliefs that form the core for what we have called our "interpretive experience."

An animal, for example, reveals the stress response when it is physically hurt or when its actions are actually blocked so that it truly is not in control. An animal does not appear to analyze, fantasize or worry in advance as humans do. As humans, we learn to analyze, fantasize and worry because early in life we realize that we are fragile creatures. It does not take long for a child to note that lack of food or the wrong food causes abdominal distress, as does overeating; that sharp edges can cut one's flesh as well as cause pain; that one can get bruised if we hit something too hard or something hits us too hard; that bones can break and that fire can burn.

To prevent possible injury, a child is repeatedly told by his parents and others not to touch fire or he will be burned; to keep away from knives or he might be cut and bleed; to stay away from rocks or he might wind up with a bruised body or broken bones; to watch out for cars and bicycles or he might be crushed; to be very, very careful because besides pain and discomfort, one can also die. Soon the child lives by a tremendous number of negative injunctions: don't touch that; don't eat that; don't go there; don't do that.

Little by little, a child learns to operate to some degree out of fear; fear of being hurt, scolded, punished, deprived or possibly even killed. Soon the average child begins to feel some degree of vulnerability even if he projects an outward appearance of calm and contentment.

This feeling of vulnerability often leads him to respond defensively. Some children become somewhat shy and introverted, building a wall around themselves, afraid to meet life on its terms, for fear of being hurt. Others become more daring or aggressive in order to stand out a little and prove their worth, mainly to themselves. They might begin to impose on others. They might also begin to lie, steal and even hurt others. Rather than leading to an enhanced self-image, this often has a rebound effect as the child soon realizes that whatever he does to others can also be done back to him as well. If he throws rocks, rocks can be thrown at him. If he steals, his possessions can also be stolen. If he bullies, he can be bullied. This realization only increases the feeling of vulnerability.

In addition, almost from day one, parents and society carefully teach the child certain tenets in word as well as by deed and these also become part of the child's rigid belief system. The child is repeatedly taught, among other things, that:

The individual has little or no power. He is a victim of events such as wars, epidemics, the economy, unfair laws and even of his own subconscious.

Birth and life are basically accidental. One is alive by pure chance.

The body is vulnerable to injury, disease, attack and the elements.

The individual is of little importance to nature as compared to the species.

Only the fittest survive.

The act of birth is traumatic and casts a negative effect throughout life.

Feelings are not to be trusted. They lead one astray.

People are not be trusted. They lie and steal.

People are at the mercy of their genes. If you do not like who or what you are, blame your parents.

Life is but a dream. It is fleeting and often unreal.

Suffering is part of life. One has to brace oneself against trouble and pain; grin and bear it.

Might is right. Do not be a sissy. Fight back.

Winning is the name of the game. Be a winner.

Wealth is power. One cannot afford to be poor.

Happiness is success. One must not be a loser. Therefore, "Try harder"; "Be Number One."

The world is unsafe. Bacteria and viruses can cause disease. Insects can cause distress. Blizzards, tornadoes, earthquakes, volcanoes, hurricanes, lightning, floods and drought can cause torment. And, society, industry and technology, through their stress on power and profit, constantly conspire against the individual.

With these commonly taught tenets and dogmas as guides, it is not strange that many individuals, to varying degrees, soon begin to build belief systems that are based on not being in total control. They begin to doubt themselves and their ability to affect the outcome of situations. They begin to feel that they are basically alone, that things just happen to them. They talk more about the need for "luck." This overall attitude becomes even more accentuated when parents stress obedience and demonstrate love and affection for the child mainly when the child has done something that pleases them. If love and affection are not given simply for being oneself, the child subtly and silently develops feelings of lowered self-esteem. Then it becomes not the child and the world but the child against the world.

However, as much as we would love to blame parents, society and life itself for all of this, some of the blame rests directly with the individual himself. The world is. Existence is. How we view it is open-ended, a product of our own choosing, of the choices we make. One can choose to see a cup as either half-filled or half-empty. Similarly, we can choose to judge events as either good or bad. We can choose to be happy or sad. Once we choose a view of the world or of a particular event, we tend to hold on to it tenaciously and to believe it. Believing in it makes us begin to experience it literally the way we view it. This further convinces us that our view has to be true since this is what we are experiencing. Yet, without doubt, this view is of our own creation. It is what we have chosen to "see." It is not "Vision." It is our vision of "Vision." And, this type of choosing, what and how we perceive, starts early in life and becomes conditioned very easily.

Therefore, when anyone (a child or an adult) "chooses," so to speak, to feel powerless and not in control, when an individual "chooses" not to trust fully in his own powers and capabilities, he then loses the ability to make consistently effective or appropriate choices. This self-created limitation only tends to make him feel even more powerless. It is a vicious cycle that usually ends with anger and fear directed either onto the world or back onto himself.

In essence, then, we create an environment, a world, a society, a reality for ourselves. It may not always be one that is in total harmony with nature and the natural order of things. To the degree that it is not, and to the degree that we resist it, it becomes a major source of "distress." Therefore, we see again that stress, or distress, is really an indicator of a disharmony or dissonance between the individual and his environment.

Theoretically, dissonance of any kind should become the feedback that indicates some form of readjustment is required so that our behavior can be modified. Unfortunately, in practical terms, it usually does not work that way. Though we may desperately want to change, to be happier and to stop worrying, we do not usually change because we do not alter those resistant belief systems that cause the problems initially. Instead, we just tend to suffer and bemoan our fate.

Changing our basic belief systems and attitudes is extremely difficult to accomplish because, for the most part, we tend to be totally unaware of what these core beliefs and attitudes are.

More often we find that we reinforce our current behavior instead by:

1. overrelying on inner talk and images that are drawn from past experience, from our memory of what was, rather than just letting the moment exist free from prior notions and influences;

2. projecting our feelings of vulnerability into the future, thus, in effect, hindering and paralyzing our ability to move off base. We daydream about how things could be while inwardly believing that it is our fate not actually to experience life that way;

3. constantly judging ourselves and others with rigid and often negative opinions that tend to block the possibility of changing relationships and ideas about the world.

This resistance to change, to modifying one's behavior can be attributed directly to the ego. Practically speaking, the ego is the device that the mind creates and uses to establish its individuality, its separateness, its distinctiveness and its uniqueness--even if it is all one big illusion.

In visual terms, the ego can be compared to the fovea, the central mechanism of the retina of the eye that is used literally to "grasp" a visual target for meaning, discrimination, and as a base for orientation in space. Like the fovea, the ego localizes and specifies. The overall "ground of awareness" is similar to the action of the peripheral mechanisms of the retina of the eye, being all-inclusive, accepting everything for potential consideration. The ego, however, again like the fovea, excludes more than it includes. Part of what the ego includes is the need for control, the need to resist change, to preserve itself in the face of perceived threats. It is the ego that establishes and maintains the belief systems which run our lives.

If we have a belief system that holds that it is wrong to show anger or to be angry, then we might tend to retreat when active defense would be more appropriate. If we have a basic belief that one should constantly assert oneself, then we might tend to be more aggressive or become defensive even if it were more appropriate to back down.

If our ego holds two conflicting belief systems, one that says, "Assert yourself; never be satisfied with less than perfection" and another that says "Don't fight; turn the other cheek," then this ambivalence will tend to produce a more acute stress response than might ordinarily occur.

Since it is one's outlook in regard to an experience that actually determines its effect, it is less appropriate to think of a stress threshold, per se, as if the stressor itself was the major factor, than to consider the threshold as one based partially on physiological factors and, more importantly, on one's psychological perception of vulnerability.

This is one of the reasons why a given event does not always produce the same response in different individuals or even in the same individual at different times. It is not the act itself that produces the particular stress response but one's approach to the task. This goes way beyond the physical mechanics of any particular task or of the physiological operation that is necessary to perform it, or of the cognitive requirements that the task demands. The major factor rests in one's psychological approach to the task. And this psychological approach (one's personality, attitudes and belief systems) is reflected throughout all operational aspects of one's life.

Now perhaps we can understand a little better the depths of the profound relationship that A. M. Skeffington (1955a), the father of modern behavioral optometry, only touched upon when he declared that "He who is unstable in space (his visual world) is insecure in his ego."

Chapter 15

The Ego, the Mind and the Body

We have seen that what is called the act of perception is actually a process of selection and evaluation. It involves choosing an aspect of the environment to attend to in order to get meaning from it while, at the same time, giving meaning to it. Giving and getting meaning involve making judgments that affect how we view both ourselves and the external world. These judgments, in turn, are influenced by deeply held core beliefs. Beliefs of any kind are all memory based, dependent on what has been previously experienced. Perception, therefore, is past- oriented--a way that the past reaches out and influences the present. Finally, the whole process of belief systems, judgment making and appraisal which is so basic to the act of perceiving and so fundamental to the stress response, is a product of the ego.

The Ego

To Freud (1938,1949), the ego represented the conscious, logical part of the mind that acts as the individual's contact with reality. That aspect of the mind which is considered to be the "conscience" of the individual, concerned with right and wrong, Freud labeled the superego. Other psychiatric and psychological theorists have contributed different concepts and views about the ego. What follows is a pragmatic synthesis of many ideas on the subject (Jung 1953-58,1923; Edinger 1972; James 1950; Tart 1975; "A Course in Miracles"1975; Fromm & Suzuki 1960; Perls 1969; Assagioli 1976,1973; Ornstein 1972).

The ego can be considered as a device established by consciousness to localize attention around a canonical center, a base of operations. The ego is simply an aspect of the mind, of thought. It is thought using a part of itself to know itself, to know that it is thinking and perceiving, to know that it exists. As part of the thought process, the ego is geared toward making decisions and judgments. It is a "center" for analysis and for the organization of mental space.

The ego establishes itself as the "I," the "self," the "analyzer," the "observer," and the "judge" and as such becomes the mechanism through which perception selects and colors incoming sensation. All events are perceived only within the framework set by the ego. All of reality, in fact, is perceived as the ego wishes it to be viewed.

It is not unusual for a person to have more than one ego or "self," each with its own battery of memories and interpretations to draw upon. Extreme instances of this have been dramatically noted in psychiatric case studies of "multiple personalities" such as that written about in the book, The Three Faces of Eve. However, it is also possible for the average person to utilize more than one ego where the divisions between them and their effects are hardly noticeable.

The ego also tends to be divided into personality segments that come to the fore depending on the situation a person is in. One can think, act and feel like a parent. One can also be someone's child, sibling, friend, teacher, student, employer, employee or spouse. One can also be a professional, a sportsman, a politician, etc., and in each framework act, talk, and think differently. These divisions are usually called subpersonalities but in reality it is just the ego segmenting itself and playing different roles.

The ego is dependent on two major factors. The first is memory. Without previous experiences to draw upon, the "self" has no continuity. Every event is a new occurrence experienced in a new way. Without a past to draw upon, one is constantly a new individual, without a history, tradition, conscience or the ability to make judgments. Without memory, there is no time, just "now."

A second factor is verbal language. The ability to label, classify and analyze experience is a function of verbal language. So is worry, caring, hoping and making value judgments. Without language, the words I and self and all they stand for are meaningless. In fact, as Julian Jaynes (1976) points out, language was the necessary prehistoric precursor for the development of ego consciousness as we know it today.

Ego consciousness adds a very unique dimension to human life. Without an ego, we would be like the animal kingdom. Like animals it would be possible to experience pleasure, contentment, happiness, joy, ecstasy, sorrow, fear, rage, anger, hunger and thirst. The ego, however, translates each experience into "I have pleasure," "I am content," "I am joyful," "I am ecstatic," "I am sorrowful," "I am afraid," "I am upset," "I am angry," "I am hungry" and "I am thirsty."

Once the "I" is added, one begins to stress differences. The use of the "I" divides and segments oneself as an individual from others and from everything else one experiences. One becomes more of a detached observer rather than a full participator in events. In addition, saying "I am" or "you are" means that I am different from you. Being different, I must therefore be either somewhat better or worse than you, stronger or weaker than you, more or less intelligent than you, better looking or uglier than you, richer or poorer than you. This has to be because if I were equal to you, I would be you. But I am not you. I am me. I am. Therefore, the more rigid and forceful is one's ego, the more one stands apart from everyone else and the more one tends to judge others in relation to oneself and oneself in relation to others.

Without verbal judgment, there would be only primary or secondary (symbolic) experience. Without verbal judgment, based on the recall of what we selectively draw from our memory circuits, there would be less fear about what is and less worry and apprehension about what might be. By involving the past and projecting it into the future, the ego creates psychological time. Psychological, in the sense that while both recall and future fantasizing occur "now" in the present, the use of recall and future fantasizing makes us live psychologically in everything but the present.

The interaction of the ego in relation to the mind has some interesting ramifications. For example, the mind creates the ego, yet the ego acts as if it had created the mind. The mind, however, inwardly realizes that the ego is far from being all-powerful and that this impression is only an illusion that the ego likes to create. In spite of this realization, the mind often subordinates itself to the ego, tending to act as if it believed in and even feared the apparent powerfulness and superiority that the ego is prone to project. In fact, a primary source of insecurity, anxiety and fear stems directly from the mind's rejection of the awareness of its true invulnerability that is always available from its own intuitive sources of knowledge and its apparent outright surrender to the ego. It is difficult for one to be at peace with the world when there is such a conflict within oneself. It is this acquiescence and subservience on the part of the mind to the ego in its interaction with the world that is a basic source of distress.

The ego, in turn, which should feel more powerful and dominant as the mind acquiesces to it, often feels more vulnerable instead. This occurs for two reasons. First, the ego is aware, at its deepest levels, that it is simply a limited extension or projection of the mind and that other levels of consciousness exist that have the potential for greater power and control than it does. Second, in its functional opposition to these other, more intuitive levels of the mind, the ego begins to identify more with the body than the mind. As this happens, "I am" begins to represent "I am a body" or, at the least, "I am a minding body" rather than "I am a function of mind in a body." To the ego, the "I" and the body become almost indistinguishable. This identification causes the basically neutral body to begin to reflect physically and physiologically the effects of the ego's conflict with the mind. As the body physically reflects these effects, the ego tends to feel more vulnerable. It then reacts to this feeling in a greater manner than it did previously, thus making a vicious cycle out of this internal ego-mind-body "dance."

The ego has a number of strategies in its repertoire to cope with this feeling of vulnerability. It can play the "helpless" game by withdrawing within itself while considering the external world as being alien and imposing. It can strive for the feeling of superiority by becoming more aggressive and repelling what it perceives as a threat. This is accomplished by attacking one's feelings, one's intuitions, one's body, other people and other things. It can also play the "guilt" game by fostering feelings of self-guilt and self-blame or by projecting blame onto the world, trying to make others feel guilty or inadequate.

Whatever strategy the ego uses, the end result is a reinforcement of the perception of itself as separate, autonomous and powerful. Toward that end, it will try to have as many people as possible like it, support it, agree with it and cater to it. It attempts to attract things to itself. Its operational mode is taking, not giving. When it does give, it does so only to get something in return, what Selye calls altruistic egoism. In other words, whatever is done for others is done only for some form of personal gain or perhaps to place the other person in its debt. The ego may rationalize that it wants love, closeness and affection but it tends to remain sufficiently detached and aloof behaviorally to prevent this from ever being fully achieved. To truly love, requires literally transcending the influence of the ego.

Freedom from the negative effects of the ego may be a difficult goal to achieve but it is not an impossible one. It is, however, a desirable goal to strive for. Freedom from the ego means freedom from the effects of the past, freedom from psychological conditioning, freedom from fear, freedom from guilt, freedom from projecting onto others the attitudes we have about ourselves and freedom from the perception of vulnerability.

What does not work is becoming angry at ourselves when we are unhappy with the way we react to people or events. Attacking the ego or ourselves only reinforces the problem. For example, if you find that you are angry at someone and you then start saying to yourself, "That isn't right of me. I shouldn't prejudge this person. The way he is today may be completely different from the way he was in the past," your thoughts may be honorable but what you have actually done is to set up a second internal verbalizer to tell the first one that it is not thinking properly. This is, in effect, a form of attack on yourself that fosters further division. You are actually playing the ego's game by setting up another ego-type mechanism as if it were a second ego to put down the first. However, one is still judging and analyzing. The ego is still at work. One has changed the mechanics of the situation but not its basic form.

One way of more effectively freeing oneself from the bounds of the ego is simply to become aware of how it works, seeing it in action as one acts and reacts to ongoing events, without judgment. Seeing that one's responses come from often-hidden beliefs established by the ego helps in freeing oneself from the power that the ego tends to hold.

Another way of transcending the negative effects of the ego is to begin to understand the direction the ego usually takes in responding to events and then consciously thinking or doing the opposite, without anger or judgment as to right or wrong. For example:

> Since the ego stresses individuality, attempt to stress unity with others and being at one with the universe.
>
> Since the ego sets up the feeling of not truly being in control, attempt to feel totally safe, supported and protected.
>
> Since the ego likes to get but not give, attempt to give more and take less. Give without the feeling that you are doing a favor.
>
> Since the ego tends to identify with the body, attempt to identify more with the mind.
>
> Since the ego thrives on comparisons and differences, attempt to see everything as being equal aspects of a total picture. See others as being extensions of oneself. Accept differences without judgment.
>
> Since the ego acts as if it were superior to the mind that created it, attempt to realize that the ego is only a tool to be used and that it can be controlled and modified.

The Mind-Body Controversy

In highlighting the actions of the ego, we touched on its relationship to the mind. In investigating the psychological and physiological elements of the stress response, we discussed the interaction between mind and body. Without specifically mentioning it as such, in probing these different areas, we were indirectly touching on a much deeper philosophical issue that has become known as the mind-body or mind-brain controversy.

Most people would accept that there are two fundamental aspects to human life: a physical form and something less definable, usually called mind. The physical form is easy to substantiate. It can be verified by the senses. It can be seen, touched, measured and analyzed. In the case of the single cell, it can be observed through the use of a microscope which acts as an extension of the senses.

The mental aspect, however, is more difficult to observe and analyze directly. It can only be inferred by observing the actions of ourselves and others. We can also rely on our own subjective experiences. It has been through personal experience that the following has become included under the rubric of "mind": the ability to think and reason; introspection; the emotional, tonal quality called contentment, unhappiness, joy, fear, delight and rage; the internal visual, auditory, thermal, kinesthetic, gustatory, olfactory and general physiological (organic) imagery one generates; one's basic reactivity and responsiveness to external and internal stimuli or events; general awareness; and self-awareness, the awareness that it is we who look out at the world and at ourselves and judge what we see. From this personal vantage point, many have then extrapolated that the signs of responsiveness that are noted in other living things, such as the single cell with its aversion to certain stimuli and its attraction to other stimuli, are possibly more primitive attributes of the mental awareness that are observed in ourselves.

This apparent dichotomy between one's physical state and one's inner mental state has intrigued and perplexed man from his earliest beginnings. It was always easy to agree on the physical aspects and to give them specific names. The gross physical form of the human being was called the "body." Its parts were all individually labeled, i.e., the arms, the legs, the stomach, the head, the eyes, the ears, the brain, etc. The mental aspect was given many names, the definitions of which were and still are far from uniformly agreed upon. Some of the names given to the mental aspect are mind, awareness, consciousness, affect (emotion), thought (cognition), energy, spirit and soul. Over the years, these two aspects, the physical and the mental, have been the source of much lively debate that has been carried right into modern times.

The main issue that has perplexed generations of thinkers is whether or not the mind and its attributes (consciousness, awareness, etc.) are functions of the body (or brain).

From early times, mind had been thought of as being fundamentally different from the brain and the body. It was accepted that each operated under separate sets of rules and principles even though they obviously interacted with each other. Philosophically, this concept became known as dualism.

Plato was the father of Western dualism but it was Descartes ([1637] 1958) in the seventeenth century who set it in its modern philosophical context by postulating that the body was matter and the mind spiritual and that the mind could exist theoretically without the body even though, in the living human being, they could and did interact. Gilbert Ryle (1949) called this concept the "Ghost in the Machine."

The seeds of opposition to the philosophy of dualism were also formulated in the seventeenth century around voices such as that of John Locke (1965) who postulated that all knowledge came from sensory observations and that the mind (or spirit) started out as a tabula rasa (a blank slate) and was incapable of having any ideas without the physical senses as a source.

Since that time, especially in this century, there has been a growing rejection of dualism in favor of a monistic philosophy which stresses the belief that the mind is a direct product of the neurochemistry of brain function. Monists assume that if the mechanisms behind mind function are not as yet known, they will be, once further discoveries are made. Monists are generally materialists in that they believe that all human function, mental or physical, is potentially capable of being explained on a mechanical basis (Rosenthal 1971). In fact, many materialists question the "Ghost in the Machine" concept by asking what the nature of the "Ghost" is that controls the "Machine" and what kind of "Ghost" controls it, etc. This type of questioning involves an infinite regression and indicates that the "Ghost" (representing mental states) is considered as being basically mechanical or neurochemical in nature.

Many researchers and theorists today tend to be monistic. Maranto (1974), for example, quotes Vernon Mountcastle, a neuroscientist at Johns Hopkins University, as stating that "old notions about the separateness of mind and brain were being swiftly supplanted by a new view that linked the two inextricably." Maranto also quotes Floyd Bloom, a neuroanatomist at the Scripps Clinic in La Jolla, California, as saying that "scientists will someday be able to add together information about the brain's circuits to show that properties of the mind are a consequence of brain activity."

Anderson and McClearey (1984) offer a compromise approach. They propose that consciousness is not something given to the human being at birth, but instead is an emergent function of the brain and the body that is achieved by development. However, "once it has emerged, it becomes relatively independent," being capable of modifying and even overriding brain and body processes.

There is, however, a small but growing group of dissenters, individuals who are prominent scientific thinkers, who question the materialist monist philosophy and find it less than adequate. For example, Sir Arthur Sherrington (1947), the noted neurophysiologist and philosopher, felt that the concept "that our being should consist of two fundamental elements offers...no greater inherent probability than that it should rest on one only."

Wilder Penfield (1975), one of the most noted of modern brain researchers, worked as a scientist for over fifty years trying to prove that mind or consciousness was a direct function of the brain. After all that time he felt compelled to conclude that "there is no good evidence, in spite of new methods...that the brain alone can carry out the work that the mind does." He also added "that it is easier to rationalize man's being on the basis of two elements than on the basis of one."

R. Buckminster Fuller differentiated between mind and brain very distinctly. To Fuller, the brain as part of man's physical equipment is the repository of memory, but the mind and mind alone is what reviews the total inventory of experiences (Snyder 1980).

Konrad Lorenz (1977), the father of modern ethology, has stated that "the autonomy of personal experience and its laws cannot in principle be explained in terms of chemical and physical laws or neurophysiological structure, however complex." He concludes that mind may be as different from brain as inorganic matter is from organic life.

Finally, the noted philosopher Karl Popper and the eminent physiologist John Eccles (1977) feel that besides physical processes and forces, there are mental states that are equally separate and real. Rather than presuming, as do the materialists, that someday research will have uncovered all the physical and brain mechanisms that are behind mental function, they assume that the enigma of the relationship between the brain and the mind is so complex that it might well be insoluble.

This controversy is more than an interesting philosophical exercise. It has very important implications in terms of everyday human behavior. Without most people realizing it, it affects each and every individual at the most fundamental level in his or her daily life. The aspect that is most important is not whether the mind and body are actually separate or whether one is derived from the other but how we perceive the relationship in ourselves at an ego level, specifically what we perceive ourselves to be. It is this distinction that is basic to the stress response.

There are two extremes on this mind-body, self-perception continuum. At one extreme there are those who perceive themselves as being primarily a physical body. At the other extreme there are those who perceive themselves as being primarily mental or spiritual organisms. Of course, every few people are extremists in their beliefs. What is important is that different self-perceptions are reflected in different behavioral outcomes.

Those with "body" orientation in their perception tend to view their mental states as attributes of physiological function. They see all of nature as having a mechanical basis. They view the body as a mechanical instrument, often comparing the brain to a computer and the eye to a camera. They identify with the body, realizing that it is fragile, easily injured and destructible. They conceive of life as being finite in nature, programmed for inevitable destruction, as all mechanical things are, and as being just an accidental convergence of the right atoms, molecules and chemicals at the right time and place. Since living things are solid, mechanical,limited and space occupying; since life is of relatively short duration; since one does not ask to be born or know when the end will come; since bodies are fragile, the individual tends to feel in competition, pitted against other individuals and against the world. It is this "body" orientation on the part of the ego that maximizes the perception of individuality, division, segmentation and separation.

Those who perceive themselves mainly as a "body" tend to feel more vulnerable. They constantly tend to compare themselves to others and therefore are more prone to be hurt emotionally. They see themselves as being too short, too tall, too thin, too fat, too weak, too muscular, too ugly, too handsome, too serious, too jocular, too lonely, too friendly, too poor, too rich, too ignorant, too educated, too light-skinned or too dark-skinned. They identify with time-space which is the medium in which bodies or solid things operate. They see themselves as prey to genetic defects, environmental deprivations and diseases of all kinds. They are often atheists or agnostics. When they do profess belief in a deity, they usually "create" their God in their own image. In a male society, He tends to be male. In a white society, He is perceived as white. He is humanlike in nature, almost always anthropomorphic, capable of manifesting the positive and negative emotions that a human being exhibits except in cosmic proportions. Body-oriented people create their God in their own image and then compete with what they have created in order to feel more in control and to be something more to themselves than just a body. They judge themselves, others and events as they feel the God they created judges them. They are highly intent on controlling events and even themselves to prove their self-worth since they perceive themselves to be in competition with all other bodies. They may even do things to hurt themselves or get themselves sick to prove they are in control. They are often sensitive and react defensively since, after all, bodies are fragile and vulnerable.

On the other extreme are the mind-oriented individuals who feel that the essence of life is mind, spirit or soul. They feel that consciousness pervades the universe and that they and their companions in life are part of the Universal Mind, a greater Consciousness; that all minds, though separate, are interrelated. They perceive themselves as being at one with the universe and with each other. They feel totally supported by the Universe, Life, Consciousness, Mind or God. They cannot conceive of the major component of their life, or life in general, as being finite or destructible. They cannot conceive of the essence of life being rigidly segmented, totally competitive and a survival of the fittest. They do not reject the existence of the body but usually consider it to be either a learning device for the mind or a device through which they can conveniently forget their origins in order to act out a part in a real-life "play" in which the world, as we know it, is their stage.

Mind-oriented individuals see through the structural differences of the physical body, its size, strength, looks and color to the consciousness behind it. They identify with the commonalities found among all life--the awareness, concern, thought and emotion that lie beneath and behind one's actions. They believe that their limits are the limits set by their own beliefs. They agree with Franklin D. Roosevelt in that "the only thing there is to fear is fear itself," that self-created terror that comes from the lack of deep faith is part of the total identification of mind with Mind in a way that body could never be identified with Body or body with Mind. This type of identification fosters a feeling of being invulnerable in regard to the aspects that really count in life and living.

In many different ways and from many different perspectives therefore, we have seen that the stress response, primarily distress, relates directly to the perception of control. The perception of control, in turn, relates to the perception of one's vulnerability. And, one's feeling of vulnerability is directly related to one's core beliefs and to the degree to which the ego perceives itself as a fragile body constantly having to defend itself from a perceived, hostile world.

It should be realized that the intellectualization of one's self-perception may not match one's deeply felt emotion. For example, just verbalizing that one believes that the mind is more important than the body does not necessarily mean that this is one's core belief. It is comparatively easy to rationalize the apparent logic of a situation while responding to an inner belief that lies hidden even from oneself. Therefore, it is not what a person professes that reveals his beliefs as much as what he does, the way he does it and how he feels doing it. This means that the best way to discover what anyone, including oneself, deeply believes is simply to observe and become aware of one's reactions to life events rather than listening to the words that are uttered.

Chapter 16

Minimizing Stress--I

We have now come to a critical point. If the stress response, specifically distress, is primarily due to one's attitude about an event, the choices one makes as to how to perceive a situation, and the way it is experienced, then what can be done about it? How can we prevent stress? If we cannot prevent it, how can it be minimized? How can we change our attitudes and belief systems?

Some pertinent suggestions have already been outlined in various sections of this book. In this chapter we will present key recommendations on stress reduction as found in the current professional and lay literature starting with Hans Selye.

Selye's Advice

Selye ([1956] 1976,1974b,1978,1970) has a series of ideas on stress reduction. He admits though that many of his conclusions are based on his own personal life experiences and observations and that they "are only superficially, or not at all, related" to his stress syndrome research (Selye 1974b). Many researchers also take this liberty when they feel the need to show that the results of their work have some practical relevance. At least Selye openly acknowledges when he makes these practical leaps. Not all do.

Selye's basic research observations demonstrate that stress reactions can become stereotyped so that the same exaggerated neuroendocrine response occurs whether it is appropriate to the situation or not. His research as well as his life experiences have indicated that conditioned stress responses can be altered in one of two ways. First, by exposing the individual to a different and more intense stressor, as a form of "shock" or "disruptive" therapy, and second, by providing complete rest to give the system time to "forget" its conditioned stereotyped response to the stressor.

In determining exactly what to recommend in each instance, Selye suggests viewing total stress as a relationship between local and general stress, specifically as a quotient based on the following formula:

$$\text{Stress Quotient} = \frac{\text{local stress in any one part}}{\text{total stress in the body}}$$

If one is experiencing excessive local stress, one should seek diversion; do something different, not necessarily something more relaxing, just different. It is a method of putting something in the way of one's worrying thoughts.

If there is too much stress on the body as a whole, if one is under excessive general stress, one should seek rest, not diversion. Just get away from it all for awhile.

Selye offers the following suggestions as well:

1. Since eustress (positive stress) places less of a demand on the body, Selye suggests doings things that are pleasureful rather than an obligation. He stresses this point by quoting an old Austrian peasant woman's ditty:

Imitate the sundial's ways
Count only the pleasant days.

2. Selye also suggests that one should get to know oneself better in order to seek out one's own natural stress level. If you discover that you are a "racehorse" then do not live like a "turtle." If you discover that you are a "turtle" then do not live like a "racehorse."

Along the same line, knowing oneself also allows one to know what causes negative effects. Since "knowing what hurts you has an inherent curative value," Selye passes on Socrates' advice to "know thyself" and Matthew Arnold's suggestion:

Resolve to be thyself and know that he
Who finds himself, loses his misery.

3. Selye also suggests that one's goals be chosen carefully, making sure they are one's own goals and not those that are being imposed by others.

4. He recommends looking out for oneself by giving of oneself thereby making one necessary to others. In this way, one can gain goodwill, respect, esteem and gratitude, and by so doing never be alone. Since this outward giving is done for a form of personal gain, Selye called it "altruistic egoism."

5. Selye's primary rule, however, is "Don't fight over every little thing just to be right." If you are going to struggle, do it when it is obvious you can win. If you sense you are in a losing situation, give up.

"Fight always for the highest attainable aim;
But never put up resistance in vain."

These are Selye's conclusions. They are pragmatic and useful. Others have made slightly different suggestions.

Other Suggestions

Tubesing (1982) recommends the following:

1. Reorganize: Set goals and work toward them.

2. Change scene: Find or create a setting conducive to inner peace. Tune in to the feelings of others.
3. Change your mind: See the positive elements in every problem. Accept the incongruities of life and learn to laugh at them. Praise rather than criticize yourself.
4. Build up your strength: Eat a healthful diet, arrange daily relaxation times, get in the habit of exercising.

To overcome "career burnout" Hensley (1984) recommends:

1. Get adequate rest.
2. Get adequate exercise.
3. Make positive use of the past. Keep and read a log of your most productive days.
4. Make positive use of the future. Always plan something to look forward to and be excited about.
5. Study something new for mental stimulation.
6. Learn to say "No."
7. Travel more.
8. Develop a hobby.
9. Set realistic and flexible goals.
10. Set up a support group of friends and peers.

In a similar vein, to break the tension of work, Wingerson (1984) recommends:

1. Don't make lists of things to do. If you must make a list, stick to it without letting extraneous things become added to it.
2. Don't eat at your desk or take business lunches. Have lunch with someone outside your profession and don't talk about work.
3. Force yourself to hurry less. Don't wear a watch. Don't leave the table as soon as you have finished eating.
4. If you find yourself worrying about time or about success, say "Stop" to yourself out loud and then think about something relaxing.

Shaffer (1982) recommends the following in order to develop stress resistance:

1. Decide to live--embrace life at home, at work, or at play.
2. Have a will to live and have specific reasons for being.
3. Determine how you can live well and thrive.

4. Have an optimistic attitude about your life and yourself. Expect things to turn out well even in the face of adversity. Expect success and be confident in your abilities.
5. View yourself as the master of your fate. You control your own destiny.
6. Practice an open learning mode. Think divergently and creatively.
7. View stress and changes as challenges and opportunities to learn and grow.
8. Practice the solution-finding dialogue daily. Step back, gain an overall perspective, define the problem and then find solutions. Don't think negative thoughts.
9. Become an active participant in your destiny. Make a balanced commitment to family, work and play.
10. Live responsibly. Accept and seek all types of feedback. Strive to improve yourself.

Meyer (1981) notes how words reflect one's self-image and recommends the following:

1. Change "should" to "could." Should implies that someone else is telling you what to do. Could implies you have a choice.
2. Change "but" to "and." But erases whatever was said before. And is a connecting word.
3. Change "can't" to "won't." Can't is the perfect excuse for avoiding responsibility. Won't conveys a sense of disarming honesty and deliberate decision,
4. Change "try to" to "will/won't." Try to is a put-off phrase. It is, in fact, a decision not to do something. Will and won't are words of decision and commitment.
5. Change "have to" to "want to." Have to implies an irrevocable command to be unhesitatingly carried out, enticing us to rebel. Want to implies an act of personal willingness, responsibility and freedom of choice.

Friedman and Rosenman (1974) recommend the following as an antidote to being a Type A person:

1. Stop hurrying. Leave more time to do things. Don't leave things for the last minute. Don't try to do too many things at once.
2. Work in an environment that promotes peace.

3. Listen more. Let others do the talking.
4. Meet yourself at lunch. Let lunch be break and not a continuation of work.
5. Don't be a deadline worker. Don't feel you must finish everything. Don't be under pressure to finish by a given time.
6. Search for aloneness. Find time each week to be alone where you can appraise your life, its problems and goals.
7. Don't be hostile. Use your sense of humor to defuse your anger.
8. Don't be goal-oriented. Enjoy today's pleasures.
9. Don't do everything yourself. Delegate authority.
10. Refine your character. Work to attain things worth "being" than "having." Maintain your spiritual and intellectual faculties. View myths, ritual and tradition as being a delight rather than an insufferable obligation and waste of time.
11. Establish reasonable life goals both for your career and private life.

Friedman and Ulmer (1984) list the following as the ten freedoms from Type A behavior:

1. The freedom to overcome your insecurity and regain your self- esteem.
2. The freedom to receive love.
3. The freedom to mature.
4. The freedom to restore and enrich your personality.
5. The freedom to overcome and replace old hurtful habits with new life-enhancing ones.
6. The freedom to take pleasure in the experiences of your friends and family members.
7. The freedom to recall your past life frequently and with satisfaction.
8. The freedom to listen.
9. The freedom to play.
10. The freedom to enjoy tranquillity.

To Robert Eliot (Eliot & Brio 1984) "Rule Number 1 is don't sweat the small stuff. Rule Number 2 is it's all small stuff, and if you can't fight and you can't flee, flow..."

Eliot also feels that much of one's stress comes from the internal conversations we have with ourselves. He lists the following ten common irrational self-talks and their antidotes taken from the stress-reduction workbook, Becoming Thick-Skinned, by Hermann Witte.

1. "Things upset me." Change to "I upset me," thereby accepting responsibility for one's behavior.
2. "I have no control over the nature, intensity and duration of my emotional responses." Change to "I may not be able to control getting upset, but I have significant control over the intensity and duration of my feelings."
3. "My emotional and stress responses are inherited." Change to "My feelings and behaviors are learned and if they are ineffective, they can be unlearned."
4. "I can't change the way I react." Change to "I can change my emotional responses to stress through realistic constructive self-talks."
5. "Things should and must go as I demand." Change to "It would be great if things go the way I want, but I am prepared to deal with whatever happens."
6. "I can't believe it. I don't understand it." Change to "I can understand my sources of stress if I put my mind to it."
7. "It's horrible. I can't stand it." Change to "Things are never as bad as I can make them seem."
8. "My shortcomings make me a bad person, a loser, a failure." Change to "I can accept what I am even if I have shortcomings in certain situations."
9. "Others think as little of me as I do of myself." Change to "Others will accept me as I present myself and I can present myself in the best possible light."
10. "It's hopeless. Things will never get any better." Change to "There is always hope, because in the long run things will look different--especially if I can learn how to cope better with them."

It is recommended that in using these antidotes, "Act as if you believe what you are thinking and ignore the old feelings." Over a period of time, practicing the new thinking and "acting- as-if" it were so will make it come to feel right and natural.

Progressive Relaxation

Many of the programs that have been developed for stress reduction utilize relaxation procedures. These include meditation, hypnotic induction, autogenic therapy and a technique originated by Edmund Jacobson (1929,1957,1964) called "progressive relaxation." Progressive relaxation is based on the fact that it is so difficult to relax since we have become so used to tension and stress that we don't realize we are tense or under stress. Jacobson's idea was that in order to know what relaxation or a state of "no tension" was like, one first had to experience tension. He found that the best way to do this was to consciously self-generate more tension. Over the years, there have been many variations but the basic procedure is to tense a given set of muscles as hard as one can and hold this tension for a number of seconds. Then the muscles are allowed to relax and the tensing-relaxing cycle is repeated. Many individual muscle groups can be tensed in this way. One can do this by clenching the fists, hyperextending the wrist, stiffening the muscles in both arms, tightening the forehead, squinching the eyelids, tightening the muscles of the legs, abdomen and chest, etc. Jacobson suggested tensing each muscle group three times with a five-minute relaxation period at the end of the series after which the series is repeated. The series itself is done three times followed by a break of at least fifteen minutes. Some of the current variations may spend more or less time on the technique or utilize more or fewer muscle groups. However, regardless of what muscle groups are tensed or how often it is done, it is important for the individual doing the technique to become aware of all the sensations that are experienced, the feelings of tension and the feelings of relaxation.

Jacobson made two important observations. The first was that anxiety is lessened and eliminated as body muscles become more fully relaxed. Second was the fact that his technique caused relaxation to generalize to muscle groups that were not specifically worked on including the involuntary musculature.

Autogenic Therapy

Most formal hypnotic induction routines provide instructions to the subject to relax different muscle groups either starting with the forehead and ending with the toes or the other way around. Subjects to be hypnotized are asked to sit quietly, with eyes closed and with breathing slow and relaxed. As subjects become more relaxed and as hypnosis is induced, many report that they have a sensation of warmth either in the body as a whole or in the extremities. This is often accompanied by a feeling of heaviness. These observations were first reported by Oskar Vogt, a German brain physiologist, between the years 1890 to 1900.

In the 1920s, Schultz (Schultz & Luthe 1959), a Berlin psychiatrist and neurologist, felt that the sensation of heaviness and warmth was the key that differentiated the normal from the hypnotic state. He decided to stress these properties through the use of passive concentration on specific verbal formulas. Since at that time anything to do with hypnosis was viewed with suspicion, Schultz called his technique "autogenic therapy" (Luthe 1963, Lindemann 1973).

The main autogenic therapy procedure involves the use of six formulas relating to arm heaviness, arm warmth, calm heart beat, calm breathing, warmth in the abdomen and coolness in the forehead. The individual sits or lies comfortably in a quiet, dimly lit room. He begins by repeating in his mind "My right arm is very heavy." Left handers use the left arm. This is repeated six times and is followed by the tranquilizing formula, "I am completely calm," said once. This combination is done three times. During the procedure the subject should visualize himself carrying a heavy weight. The feeling of heaviness is an indication of muscle relaxation.

The second formula is "My right (or left) arm is very warm." This is repeated in the same manner as the first formula. The feeling of warmth indicates dilation of the peripheral blood vessels. This formula is used until the feeling of warmth spreads to all the other limbs. Then, the third formula is used, "My heart beat is calm and regular." The fourth is "My breathing is calm and regular." The fifth formula is "My abdomen (solar plexus) is warm." The sixth is "My forehead is cool." Schultz found that the average time for achievement of each exercise was from two to eight weeks. Total efficiency in all six procedures was usually expected in four to ten months.

For those mastering these six formulas, Schultz offered a second, more meditative, series that focused mainly on mental functions. This series concentrates mainly on the use of visual imagery, stressing the visualizing of colors, objects, abstract concepts, and other people and finally, getting intuitive answers from the unconscious.

Meditation

Many activities go under the name of meditation (Naranjo & Ornstein 1971, Maupin 1969, Carrington 1977, LeShan 1974). In essence, meditation is a passive but total form of contemplation. Most people tend to experience this mood spontaneously at one time or another. It can also be brought about by concentrating on something in particular such as a repeated word,breathing, body sensations, falling rain or a flickering candle. It can also be accomplished by being open to all sensations and experiences, internal and external. This latter approach is called "Insight

Meditation." The most popular form today, "Transcendental Meditation" (TM), utilizes the repetition of specific and secret Sanskrit words assigned to each individual depending on his individual nature. Meditation accomplishes a number of things. Most importantly, it induces relaxation. It also tends to reduce internal verbalization, the internal "chatter" that distracts one from observing and that encourages making judgments. Through meditation, one begins to observe more in general. Meditation has also been found to lessen migraine headaches, hypertension and other stress-related illnesses (Wallace & Benson 1972, Hassett 1978).

The Maharishi Mahesh Yogi, founder of TM, has claimed that thousands of secret mantras are used, supposedly corresponding to the energy impulses of the individual. Some disaffected former TM trainers, however, have revealed that the following mantras actually are assigned according to age: eng (0-11 years); em (12-13 years); enga (14-15 years); ema (16-17 years); aeng (18-19 years); aem (20-21 years); aenga (22-23 years); aema (24-25 years); shiring (26-29 years); shirim (30-34 years); hiring (35-39 years); hirim (40-44 years); kiring (45-49 years); kirim (50-54 years); sham (55-59 years); shama (60 years and over) ("Transcendental Truth" 1984).

Herbert Benson (1979) was involved in early studies relating physiological changes to TM. Feeling that the arbitrary assignment of special mantras made the approach have too much of a mystical quality that really was not needed, he recommended a secular version that has become known as the "Relaxation Response," in which an individual chooses and repeats a sound, phrase, prayer or mantra that he likes or that is meaningful to him. He can be equally successful using a completely noncultic word such as the word one.

Benson requires four basic elements in his approach; a quiet environment, a comfortable position, a verbal mantra or an object to concentrate on, and a passive attitude. The individual sits in a comfortable position with eyes closed and then deeply relaxes all the muscles of the body, starting with the feet and working upwards to the face. The subject breathes through the nose, being aware of the breathing, while saying the word "one" (or his own mantra or prayer) silently with each exhalation. This is continued for ten to twenty minutes. Benson believes that a positive belief system coupled with the Relaxation Response is a potent force for healing (Benson & Proctor 1984, Hoffman et al. 1982).

Biofeedback

The term "biofeedback" is a contraction of the words "biological feedback." It represents having an awareness of one's internal functions so that they can be controlled and modified. Those activities that operate at more conscious levels simply require awareness training to bring their actions to the attention of the individual so that control may be gained. Those biological activities, such as autonomic processes, which are generally beneath conscious awareness require specific amplifying devices to act as extensions of the senses to bring them to conscious attention.

There are several types of these devices. Most display the results either in lights or sounds. These include electromyographic (EMG) instruments which amplify muscle tension, skin temperature biofeedback instruments which measure skin temperature and indirectly reveal whether the peripheral blood vessels are dilated or constricted, electroencephalographic (EEG) instruments which reveal brain activity of different types, galvanic skin response (GSR) devices which reveal the electrical characteristics of the skin, as well as blood pressure and heart rate devices.

The role of biofeedback goes beyond just becoming aware of internal body functions and gaining simple control of them. It has become a dominant treatment procedure for many types of psychological and physiological activities. Its major role, however, has been in fostering relaxation which plays such an important part in the stress response (Brown 1977, Green & Green 1977, Tarler-Benlolo 1978).

QR: The Quieting Reflex

Charles Stroebel (1982) feels that self-regulation techniques such as TM, the Relaxation Response and biofeedback have poor transfer effects and also low compliance. People do not tend to continue them as they should and they are only effective when done repeatedly. Some do them properly but have the belief system that it is the technique that does the job. Many meditators go into a light stage of sleep while concentrating on their mantras. In addition, most people do not know how to relax and stop worrying while others get tense in the process of doing the procedures. They try too hard.

Stroebel feels that stress can be effectively counteracted by a six-second exercise he calls the "Quieting Reflex" or "QR", which should be practiced at least fifty to sixty times a day and whenever one feels tense or annoyed at something. The technique should be practiced consciously until it becomes automatic. This usually takes between four to six months for adults.

The technique itself involves smiling inwardly with the mouth and eyes, then saying to oneself, "Alert amused mind, calm body." Next, a deep, easy breath is taken that goes all the way down to the abdomen. While exhaling, one lets the jaw, tongue, and shoulders go limp, feeling a wave of heaviness and warmth right down to the toes. Then normal activity can be resumed.

As supplementary procedures to enhance the Quieting Reflex, Stroebel recommends a series of experiential training exercises that encompass meditative, autogenic therapy and relaxation principles. In addition, he recommends a nightly "Bedtime Review" to clean away the debris of the day. This entails doing some preliminary relaxation procedures and then visualizing the activities of the day on a mental "movie screen" running backwards, like a movie in reverse. This passing review should be passively observed without annoyance or being judgmental.

Stroebel also recommends keeping a "QR Diary Scan" to heighten awareness of the stresses in one's life. This involves recalling and rating seven factors that reflect states of mental or body tension. These include environmental pressures, annoyances or accidents, time pressures, emotional frustrations, fatigue, body tension and physical symptoms.

CT: Confrontation Therapy

Many of the techniques used for relaxation or stress reduction involve visual imagery, using the imagination to picture in one's mind the effects that one is trying to achieve.

Richard Silvestri (1978) describes "Confrontation Therapy" or "CT" as a process or technique that enables a person to confront a distressing emotion and extinguish that emotion through the use of imagery. The imagination is used to recreate and intensify an emotion so that it can be dealt with as a concrete entity since the best way of rooting out a negative emotion such as fear, anxiety or depression is to confront it.

The technique is composed of five basic steps:

1. Preparation: Be seated in a comfortable position in a pleasant, familiar area. Relax and close your eyes while being open and receptive.
2. Imagination: In your mind, play over the scene that evokes the negative emotion, imagining all the circumstances leading up to it and surrounding it.
3. Repetition: Image the scene again, being more detailed. Be aware of any avoidance reactions that may be present. Repeat the scene until you no longer feel uncomfortable going through it.

4. Maximization: Follow the lead of your imagination and emotions while interjecting remarks and behavior that might occur if things were brought to their extreme conclusion. Then stop.
5. Relaxation: Think of an unrelated pleasant and calming scene and put yourself in it as completely as possible for a minute or so.

The principle behind Confrontation Therapy is that the repetition of imaginary behavior that is maximized in the absence of any real consequences will lead to extinction of the behavior. It is, in effect, a structured form of Frankl's (1960) "paradoxical intention."

Two factors are necessary for successfully extinguishing an emotion. The first is that the avoidance responses must be extinguished at the same time, and second, that the subject proceeds gently with the procedure to avoid creating any avoidance responses to begin with.

Stress Inoculation

Stress inoculation is a term used for a cognitive approach to stress management (Metchenbaum 1975). The term itself refers to a process of exposing an individual "to manageable quantities of stressors that arouse, but do not overwhelm" his defenses (Novaco 1979). There are three stages to the treatment.

The first stage is cognitive preparation. This consists of the person appraising the situation in question so that he knows its nature better and knows what there is to do. In effect, he cognitively prepares himself so that the impending stress does not come as a surprise.

The second stage, skill acquisition, involves the learning of cognitive and behavioral coping skills. The therapist, for example, through challenges, attempts to teach the individual how not to take things personally. The person is taught to become task-oriented while keeping his ego out of the picture. Another approach is through self-instruction, in which the individual uses appropriate self-verbalizations such as "As long as I keep my cool, I'll be in control," to keep himself calmer and in more actual control of the situation. Relaxation training, specifically the use of Jacobson's progressive relaxation techniques, is used as a supportive procedure to develop better coping skills.

The third and final stage, application training, emphasizes practice in applying what has been learned. This is done in two ways: by having the therapist provoke the individual and by the person himself testing his coping skill in real-life situations.

Chapter 17

Minimizing Stress--II

Having outlined a number of stress reduction suggestions and procedures that are found in the current professional and lay literature, let us now turn to a different, more personal, set of recommendations based on the major themes presented so far.

1. Minimize restraint whenever and wherever possible. Repression, restriction, immobilization, in fact, restraint of any kind is stressful because it is the opposite of freedom of movement. Movement is physically and biologically given. Everything moves unless it is restrained. Freedom of movement is the source of spontaneity and creativity. Unfortunately, repression, restriction, immobilization and restraint have become so much a part of our culture that they are a way of life. Often we are not aware of the restraint but when we are, we rationalize that it is a necessity for living. In this respect, it also becomes a belief system. The first suggestion, therefore, is to encourage movement. Loosen up. Relax the "body armor."

2. Avoid situations of high input overload and high input underload since these are major stress producers. Keep in mind that even though all problems involve some degree of aberrant figure-ground relationships, a reaction to a given situation is dependent primarily on the assessment that is made in regard to what is occurring. When we respond in a given way, even if it is totally inappropriate or maladaptive, there is usually a good reason for it. Nothing happens without a reason. If we underreact or overreact, it is because of the way we perceive the event and the social constraints associated with it based on our belief systems. This is one of the reasons why we can cope appropriately in one situation and not another. In other words, when a body organ or system is affected in a certain way, it is based on a logical sequence of prior events and concurrent causes, even if we are unaware of those predisposing factors. In fact, John Lilly's (1972) work with extreme sensory deprivation revealed that the effects his subjects had while in the "deep tank" used in sensory deprivation experiments were directly related to their attitudes and beliefs about life in general.

3. Keep physically healthy through proper nutrition, exercise (mobility), and adequate rest so that the body can respond better physiologically to whatever stressors it is exposed to.

There is a lot of controversy today as to what comprises good nutrition. Many differ as to the importance of vitamin supplements. Nearly all agree, however, that: natural, rather than processed, foods should be used whenever possible; the intake of animal fats as well as other oily and greasy components of the diet should be reduced; carbohydrates should be taken in complex form, as starch, rather than in the form of sugar; and the diet should include whole grains, legumes (beans or peas), fresh fruit and fresh vegetables.

Exercise, in general, has been shown to be beneficial for increased body performance and health. Aerobic exercise, which involves a sustained increase in oxygen demand, has been found to be specifically helpful to the cardiovascular system. However, doing any exercise with tension, performing it with a "vengeance," so to speak, or in competition with others, increases stress rather than minimizes it. This tends to neutralize the possible benefits of the procedure.

Good breathing habits, specifically diaphragmatic breathing, have been found to be beneficial as well. It requires less muscular effort than chest breathing and stimulates a fuller interchange of oxygen and carbon dioxide in all parts of the lungs (Nuerenberger 1982, Everly & Rosenfeld 1981). Harvey (1978) states that "diaphragmatic breathing stimulates the solar plexus and the right vagus nerve in a manner that enervates the parasympathetic nervous system thus facilitating full relaxation." In diaphragmatic breathing, the upper abdomen rises with the inhalation and falls with the exhalation.

The concept of keeping physically healthy needs to be qualified to some degree. The body has a certain built-in flexibility and plasticity. It can adjust to many deviations from the so-called norm. It seems to be able to cope in many instances with things that we consider to be negative, including less-than-adequate nutrition. If not, more of us would be in worse shape than we are. Just remember that not everyone falls prey to an epidemic.

However, when one feels divorced from nature, not in control, when too many things appear to happen unpredictably, one becomes in effect a "victim," powerless and fearful. When this attitude is then superimposed onto the effects of potential stress producers, not only does it magnify the negative effects of prolonged immobilization and prolonged mental concentration, it also magnifies the effects of inadequate nutrition. It can literally help cause the system to turn against natural forces and even natural products as in the case of so many allergic reactions. At other times, it can even turn parts of ourselves against ourselves as in many auto-immune and ulcer-type diseases.

4. "Know thyself" (to quote Socrates). Become generally more aware. Become aware of how we often tend to constrict periphery while we are engaged in central-type tasks. Become aware of what surrounds us. Become more aware of thoughts, emotions, dreams, desires and intuitions. Become more aware of our actions and those of others. Become more aware of our reactions to other people and their reactions to us. Become more aware of how we create our illusions and our belief systems. Become more aware of how these belief systems are sustained and augmented by our inner talk. Become more aware of how we desperately search for stability and instead wind up with conformity. Become more aware of how we set up rules and regulations to make things more predictable and how these rules begin to rule us. Become more aware of how all this makes us more anxious, tense, uptight and fearful. Become more aware of whether we are responding to primary experience or to symbolic or interpretive experience; are we continually coping with something that is actually happening to us or something that we are reading about, seeing on television or creating with our own inner verbalization? Become more aware of how we respond to a real danger. Do we overreact or tune out? Become more aware of how we attend and when we are inattentive.

The importance of awareness is that it counters the negative effects of those attitudes which cause an operational constriction. It is also the first necessary step for change to occur. Change does not usually occur simply because one "wills" it to. One cannot will oneself to relax, to be happy or feel in control, especially if one has the often-hidden belief that he or she is vulnerable, unworthy or guilty. It just does not work that way. That is one of the problems with self-help books. They tell you what to think and not how to change the beliefs that underlie your thinking. Being more fully aware of one's feelings, emotions, actions, and thoughts tends to allow the beliefs behind these thoughts and actions to surface and become more consciously recognized. With increasing awareness, one begins to have a different and more open perspective toward life and living. We see more choices available and change becomes easier.

Too often one becomes "figure-bound" or "focal." Being focal means seeing letters but not the space between them; seeing words but not the space between them; seeing sentences and lines and not the space between them; seeing pages and not the space around them; seeing books and not the space surrounding them.

Being focal means seeing a speaker and not the space between the speaker and objects near him; it means seeing the speaker and not the space between you and the speaker. Being focal means not seeing fully. Being focal means a trade-off between detailed knowledge of one aspect and knowledge of the whole. Being focal means seeing what you elect to see and not what is. Being focal means thinking about something rather than experiencing the entire event. Being focal is hearing the word but not the meaning; getting the meaning but not the context; getting the context but not the premise on which it is based. Being focal is seeing the wave and not the ocean; hearing the melody and not the counterpoint; hearing the chord but not the tempo. Being focal is seeing differences but not commonalities.

The mind has the potential to take in and process a great deal of information even if, for efficiency, most of this is not consciously recognized. This lack of total recognition is a normal part of mental functioning so that the system is not overwhelmed with information. This, however, is different from the restraint and inhibition that is present in controlled, directed attention or forced concentration which limits and restricts even more potential information. When attention is forced, one may do but not attend, hear but not listen, look but not see. This is the difference between passive total awareness and the controlled attentiveness that we like to think of as awareness.

What is needed is a full and open ability to attend, what Karl Pribram has called "passive attention." Attention that is highly focused and focal is forced attention. We need to learn how to attend with a central-peripheral balance, being open and relaxed as we take in and analyze information.

The goal,then, is awareness but not forced, focused attention, observation but not compulsion. Remember that forced attention requires effort and energy. It is inhibition.

Therefore, it is not a particular task, how close it is or how prolonged it is that is the real problem. It is a problem in the sense that it may involve some degree of immobilization and may represent symbolic experience as in the case of reading. It is the approach to the task, however, that creates most of the stress. In other words, "it" does not cause the whole problem. "We" cause most of the problem. Then, by blaming "it," "we," in turn, become the problem. So, be aware. See it as it is.

5. "Let go." "Let things be." Stop judging everything. Stop giving meaning rather than gaining meaning. Be more tolerant and receptive. Enjoy and have fun. Have some trust, faith or optimism that things will work out, that the natural order is basically safe and supportive and that there is value to life and living. A lack of this kind of faith only breeds anxiety, fear, despair, paralysis of action and inappropriate overreactions.

If you were insulted by someone ten years ago or perceived that you were insulted and still carry a grudge against that individual, all your present and future interactions are going to be affected by that attitude. How much better to "let it be"-- to see the world, people, events and situations as they really are now--to let go, stop judging and stop wanting to be in control.

"Fight," we said earlier, represents attack, combatting the perceived aggressor. There are individuals who exhibit a consistent "fight" response to most situations or who always manifest an overreactive, maximum-type "fight" reaction where a lesser response would be just as appropriate, if not more so. This is brought on, in the main, by feelings of poor self-worth, helplessness and the perception of uncontrollability--in other words, a basic feeling of vulnerability.

This individual, for his own well-being, needs to "let go." But it is difficult to "let go." It has become a habit not to "let go." He often tries too hard, which is not really "letting go" at all. There is an interesting way out of this impasse, however, and that is for him to view the other party, the perceived aggressor, as being just as vulnerable, just as helpless and just as unable to control events as he himself is, if not more so. This kind of attitude stimulates some degree of understanding and compassion and also reduces the tendency to overreact and perpetuate grievances.

"Letting it be" also means staying with a problem, becoming immersed in it, going with the stress. There is nothing less effective for releasing tension than fighting it. If you try to stop blushing, you only blush more. If you "let it be," or even try to blush more, the blushing will stop sooner. If you fall into deep water and cannot swim, struggling to surface only tends to keep you under longer. "Letting go" lets your natural buoyancy take you to the surface faster.

"Letting go" and "letting it be" also mean not being influenced by that inner talker who keeps telling us what to do.

The implication in all this is that even under the most dire circumstances, we can alter how we react by how we choose to perceive events. We can perceive differently. We do not have to be a victim of fear, anxiety, anger, resentment or guilt. The choice is definitely ours.

So, again, enjoy and have fun. Flow with things. This is what eustress is all about. Remember that resistance and distress most often come from not seeing things as they really are now, from wanting to change "what is" to match some kind of created image of what we think "should be." In fact, the real source of"distress" is just this kind of misapplied thought.

If you want to find out whether you are doing the right thing or whether you are on the right track or not, try this experiment. First, ask yourself what some of your primary goals are. What means the most to you? Is it knowledge? Status? Wealth? Possessions? Peace? Then try to visualize yourself striving for each and sense which produces the least inner dissatisfaction, tension or conflict, and which the most. Dissatisfaction, tension, unhappiness and unease are all signs of inappropriate responses or outcomes. Therefore, let the feeling of tension be the real guide as to whether your responses and attitudes in a given situation are the appropriate ones for you.

It is universally recognized that many of us repeatedly choose responses or outcomes that are often inappropriate and which cause tension. Whether we realize it or not, one reason that these inappropriate responses are repeated is that they do have value. For some, they are a way to get sympathy. For some, they are away to demonstrate how strong we are and how we can endure adversity. For some, they are a way of avoiding situations. For some, they are a way of creating the illusion that we are in control. For some, they are a form of defiance to prove that if we cannot be in control of key events in our life, we can at least be in control of our own destruction. For some, they may be a form of self-punishment for feeling guilty about something.

Starting in the next chapter we will begin a discussion of vision and visual stress. But what has been said to this point has broad implications optometrically. For one thing, it implies that certain outcomes, such as myopia (nearsightedness) may come about for many different reasons. This may help explain why no two myopic individuals respond to therapy the same way. It also implies that if a person begins to recognize, not just intellectualize, that his overt problem did not just happen to him, that it represents an outcome of something more fundamental; that he is not necessarily a victim of things outside his control; that the chances are that he, in some way, created or contributed to his own problem; that he is truly in control, if he would only believe it--then this recognition alone could be a beginning step in attempting to reverse the outcome or at least, in the case of myopia, preventing its progression.

Remember the old debate about whether function alters structure or structure alters function? Well, we have just gone way beyond that question. We have introduced a third factor that is even more important than either structure or function: one's attitudes and beliefs. We can now see that one's attitudes, feelings and beliefs, even in early infancy and long before they can be verbalized, act as a "pre-set" that can alter both function as well as structure. This paradigm shift alone has tremendous implications.

Chapter 18

Vision: Going Beyond the Classical Approach

Visual stress is not a topic that can be separated from general stress theory. The visual process is a dominant function of the organism in terms of operating in the world and getting meaning from it and thus reflects the actions of the body and mind in very specific ways. For this reason a knowledge of the principles behind general stress theory is important for a better understanding of visual stress. Similarly, being aware of the principles involved in visual stress provides for a richer understanding of the general stress concept.

A Salute to Dr. A. M. Skeffington

The subject of visual stress as a major approach to understanding the actions of the visual process owes its existence to the thoughts and work of one man, A. M. Skeffington, O.D.

A. M. Skeffington was an optometrist who had a unique ability to look at things and see relationships that others overlooked. He had the ability to formulate ideas and sense the valuable from the valueless. He had the knack of posing astute and relevant questions. As an educator, he had the ability to move minds and get people to think. His was a mind that was open, attentive, aware, constructive, inquiring, analytical and discerning. It was a mind that was open for discovery, for seeing connections and for creativity.

Skeffington's work was anything but static. His ideas and concepts were constantly being modified to keep up with new research and his own expanding base of knowledge. This was especially reflected in his overall view of vision and his thoughts on visual stress. Although Skeffington's evolving model was to a great extent psychological in nature, it developed from a purely physiological base. What triggered and directed his thinking was the apparent disparity between what was observed clinically and the then accepted theoretical, conventional wisdom. There were just too many exceptions and enigmas that classical theories of vision did not appear to take into account. This became the motivation for him to look at visual function in a new and more productive way. Although our own discussions may occasionally differ or go beyond the basic Skeffington approach, much of it will owe its origin to Dr. Skeffington's seminal thinking.

Homeostatic Mechanisms

Investigators have classified the major components of the nervous system in various ways. Those considering the nervous system from an anatomical point of view tend to categorize the key components in the following manner:

1. Central Nervous System: consisting of the brain and the spinal cord.
2. Peripheral Nervous System: consisting of incoming (afferent) sensory nerves and outgoing (efferent) motor nerves.
3. Autonomic Nervous System: consisting basically of a motor system serving the maintenance of steady states within the organism, regulating body processes of respiration, circulation, digestion, etc.

From a more functional viewpoint, the nervous system can be divided into somatic or skeletal and visceral components with each of these being further subdivided into sensory (afferent) and motor (efferent) aspects. This leads to the following classification:

1. Somatic Sensory: handling impulses from the sense organs of the skin, the special senses of vision and hearing, and proprioception from muscle and joint receptors.
2. Somatic Motor: handling motor impulses to the voluntary or skeletal muscles.
3. Visceral Sensory: handling sensations from the viscera and the special senses of smell and taste.
4. Visceral Motor: handling motor impulses to the supposedly involuntary muscles of the viscera and glands. The autonomic nervous system is primarily an elaboration of this component.

Arnold Gesell (1950) divided the nervous system into three major functional components, skeletal, visceral and cortical and related each to visual function as follows:

1. Skeletal: consisting of both the sensory and motor aspects of the somatic branch of the nervous system. This division acts to "seek and hold" the visual image.
2. Visceral: consisting of both the sensory and motor aspects of the visceral and autonomic nervous system branches. This division acts to "discriminate and define" the visual image.

3. Cortical: consisting of the integrating aspects of the brain and spinal column. This division acts to "unify and interpret" the visual image.

The following extrapolations can be derived in terms of the visual process from the previous classifications. First, the retina of each eye is a somatic sensory function. Second, the muscles around the eyes that act to center on a stimulus are skeletal muscles making them a somatic motor function. Third, the muscles in each eye that act to focus for clarity and meaning are visceral musculature and are therefore a visceral motor function.

This indicates that the only sensory structure of the eye is the retina. All other aspects of the eye that are associated with the process of vision are motor. The radial and circular muscles of the iris as well as the ciliary muscle of the lens are motor components innervated specifically by the autonomic division of the visceral nervous system. The extraocular muscles associated with monocular and binocular eye movements are motor components innervated by the skeletal nervous system.

This happens to be a very unique arrangement in which the organs of sight have two major aspects of their operation controlled by different nervous system divisions and where both must work in an intimate relationship with each other for an intricate and simultaneous interaction.

The very fact that one aspect of the visual system is viscerally controlled and the other is skeletally controlled necessitates that there be some type of mechanism to protect total visual function from specific overactions or disturbances in either of the two neural divisions as they operate as a team.

The organism constantly undergoes shifts in the ascendancy of either visceral or skeletal function. The difference between sedate activities and vigorous muscular action not only involves differences in the dominance of these respective components of the nervous system, it also involves shifts within the sympathetic and parasympathetic divisions of the autonomic system itself. All these varied shifts in neurological dominance tend to affect the relationship between the visceral-motor and skeletal- motor aspects of the visual system. If there were no "play" between the functions of accommodation and convergence, the slightest change in neural dominance could radically upset the efficiency of operation of the visual process. Shifts in neural dominance are constantly occurring. They are routinely caused by changes in daily activities, thoughts, emotions, disease and stress. Since the organism has built-in homeostatic mechanisms in almost all systems to buffer the effects of external and internal change, it is only logical to assume that the visual process should have its own as well. It is this concept of homeostatic mechanisms for the process of vision that has changed the outlook on what was previously regarded as "ocular defects," "refractive errors" and "muscle imbalances."

The Ocular Defects

In the classical concept of vision, there are considered to be two major "focus" conditions, emmetropia and ametropia, and two major "muscular" conditions, orthophoria and heterophoria.

Emmetropia has been regarded as a state where the eye is in sharp focus for an object at twenty feet or beyond (optical infinity) without any need to focus or accommodate for it.

Ametropia encompasses all those conditions that do not meet the requirements of emmetropia. These have been considered as focus "defects" and have included ocular states such as hyperopia (farsightedness) and myopia (nearsightedness).

Hyperopia has been traditionally defined as a condition in which the eye is potentially capable of a sharp focus for a distance object but not without some accommodative effort. A certain amount of ciliary muscle action is needed to achieve the clear focus. According to this view, a hyperopic eye must accommodate for infinity as well as for all other distances.

Myopia has been described simply as a condition where the eye is not in sharp focus for a distance object and where it can only be brought into focus through the use of a concave lens.

Orthophoria has been defined as "perfect muscle balance, equipoise, or binocular equilibrium" (Thorington 1947). Muscle balance tests are taken under conditions of "dissociation" where fusion is made impossible to maintain. If, under these conditions, the eyes remain parallel, orthophoria is said to exist.

Heterophoria has been defined as "imperfect balance or imperfect binocular equilibrium" (Thorington 1947). It is considered to exist if, under the conditions of a fusionfree state of dissociation, the lines of sight of the two eyes are not parallel. The two major heterophoric conditions are known as esophoria and exophoria.

Esophoria is the tendency, under dissociation, for the visual fixational axes to deviate inward, for the eyes to converge or posture closer than the target of regard.

Exophoria is the tendency, under dissociation, for the visual fixational axes to deviate outward, for the eyes to diverge or posture further than the object of regard.

According to classical thinking, the normal set of eyes should be both emmetropic and orthophoric. Paradoxically, emmetropic and orthophoric eyes are the least commonly found visual condition. The most frequently noted condition has been hyperopia and exophoria. This observation led Thorington (1947) to describe the hyperopic eye as the "natural eye" or the "eye of nature." Duke- Elder (1946) also pointed out that "it would be strange if (emmetropia) was a common state of affairs, for its attainment depends on an exactitude to within a fraction of a millimeter of such measurements as the length of the eye and the shape of the cornea and the lens; and such regularity and conformity to type as optical perfection would necessitate, demand a mathematical accuracy which is nowhere realized in the constitution of living organisms. Emmetropia may be optically normal, but it is no more biologically normal than would be the universal attainment of a uniform height of five-feet-six-inches." The question that this raises is that since optical normalcy cannot necessarily be equated with biological normalcy, what ocular factors could be considered more biologically desirable? The answer to this rests in a closer investigation of the effects of neural dominance shifts on the accommodative and convergence mechanisms.

It is an accepted fact that a stimulus to either accommodation or convergence has some effect on the other function. Any factor that tends to stimulate convergence automatically causes a reflex stimulation to accommodation unless some mechanism or buffer of sorts is present to absorb this adverse effect. Similarly, any factor that stimulates the focusing mechanism automatically tends to stimulate convergence to some degree. This could result either in diplopia (double vision) or an active inhibition to prevent double vision unless a buffering device is in play. These protective buffers are the homeostatic mechanisms of vision that allow accommodation and convergence to operate together with some degree of harmony.

Skeffington first contributed the notion that hyperopia was a range that acted to protect the focusing mechanism from stimulations from convergence. In a similar manner, he viewed exophoria as a range that protected the convergence mechanism from stimulations from accommodation. He saw these visual states as being the homeostatic mechanisms that protected visual performance when shifts in dominance of either the skeletal or autonomic nervous systems, regardless of the specific cause of the shift, tended to affect adversely the intricate functioning that gives the organism clear, single, comfortable and meaningful vision.

The terms hyperopia and exophoria, therefore, came to represent two different concepts. In the classical optical model, hyperopia was a deviancy. To Skeffington, a given amount of hyperopia (.75 diopter) was considered to be the norm. In his model, measurements other than .75 diopter of hyperopia were not considered as deviances but as indications of how the organism had adapted to stress over time. Orthophoria was considered to be the norm in the optical model while exophoria was the expected finding in the Skeffington approach.

The Spring and Beam Analogy

A more concrete analogy can be used to demonstrate some of the dynamism involved in this process (Forrest 1959-60,1978). In this analogy, the structural stability and integrity of the convergence mechanisms can be likened to a beam. Attached to the beam is a spring with a weight (see Figure 22A). The weight

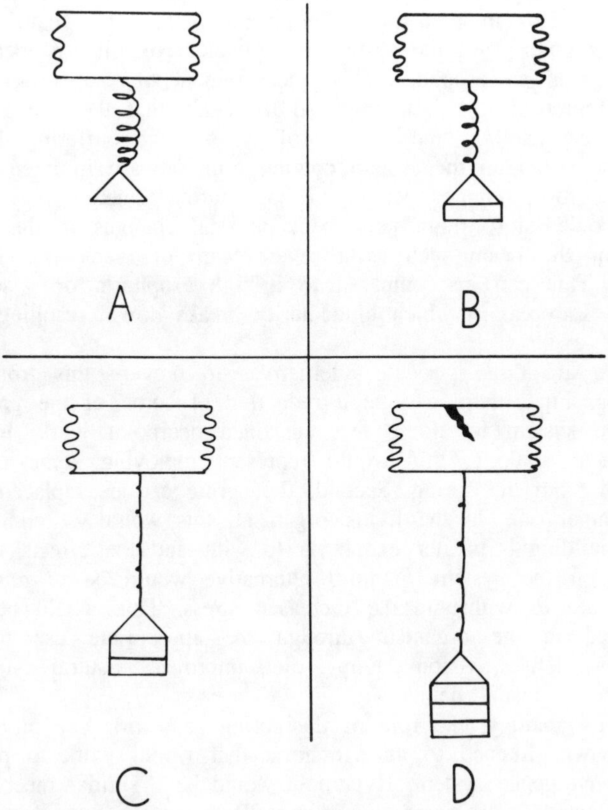

Fig. 22. Spring and beam analogy

represents the stressors affecting the convergence system. The spring represents the dynamic homeostatic mechanism which attempts to absorb the effects of the stressors so that they do not impinge on the stability of the beam.

In this illustration, the inherent ability of the spring to buffer the effect of the weight is analogous to an exophoria.

As increased stress is added to the system (see Figure 22B), the spring becomes less able to absorb all the effects of the stress. A point is reached (see Figure 22C) where the effects of further pressure result in the spring becoming taut so that any increase in tension will be transferred directly to the beam. This complete absorption of the action of the spring as a protective buffer is measurable visually as an orthophoria.

If still further tension is applied to the system, the beam itself will begin to resist the pressure with its own intrinsic strength. This internal resistance on the part of the major structural component of the system is equivalent to what is measured visually as an esophoria. The measurement of esophoria, therefore, represents the last stand, so to speak, that the system uses to protect itself in the absence of an effective buffering device.

If the stress on the system continues to increase in intensity or persists for prolonged periods of time, further changes may occur that could cause more permanent physical changes in the structure of the beam such as a severe warp or crack (see Figure 22D). This can be compared to a high esophoria or a strabismus (esotropia) in which binocularity breaks down, resulting in a turned eye.

There are three possible alternatives to prevent this from occurring. First, there can be a reduction of some of the pressure on the system by doing less sustained nearpoint work. In this particular analogy, this would represent removing some of the weight from the spring. Second, the spring can be replaced with a stronger one. In the living organism, this would be equivalent to rebuilding a greater exophoria to withstand the effects of the stress on the system. A third alternative would be to reinforce the beam to withstand the increased stress. This would be represented in the organism through the appropriate use of appropriate lenses, vision therapy, diet, improved postural dynamics and better visual habits.

In this analogy, the role of the spring as a first-line protective buffer was likened to an exophoria that would work to protect the convergence system. Hyperopia would be a similar mechanism to protect the accommodative system. Both systems (exophoria and hyperopia) must operate together as an integrated team for maximum functional efficiency.

This concept is quite different from the classical approach that sets up rigid standards of normalcy which few approach and those who do are often found to be less-efficient visual performers.

Further Changes in Concepts

F. Matthais Alexander (1974) founder of the Alexander Technique, liked to use the terms "means-whereby" and "end-gains" or "end- goals" to differentiate between the mechanics of a function and its overall higher-order purpose. A. M. Skeffington (1955b,1958a) thought the same way even though he did not use the identical words. It was this approach, however, that led him to see hyperopia and exophoria as being the means-whereby homeostatic mechanisms can function to protect the functional integrity of accommodation and convergence. It also led him to view convergence or alignment as being the means-whereby one could localize objects in space and accommodation as being the means- whereby one could get meaning from those aspects of visual space. To differentiate between the means-whereby and the end-goal, he called the higher-order function that convergence subserved centering and the higher-order function that accommodation subserved identification. Centering is the ability to know where things are in space in relation to oneself. It answers the question "Where is it?" Identification determines the meaning that the stimulus conveys. It answers the question "What is it?" In this sense they are end-goals.

Skeffington (1963) further saw that identification and centering were also means-whereby. When they interacted with the speech-auditory and the vestibular anti-gravity systems of the body, they subserved an even higher organismic function that he, at various times, labeled "vision, the emergent," "eidolon," or "iconopoesis." To Skeffington (1957d), vision in its highest and most encompassing sense emerges as a result of the learned integration of its lower-order components with other major sensorimotor modalities of the body.

Skeffington (1957d) also saw all of vision as being related to the building of a space world. He conceived of vision as essentially being the ability to retrieve "those facets of prior experience that contribute to the goal satisfaction of the moment;...the utilization of prior experiences as they relate to the immediate need" (Skeffington 1969). If one does not adequately localize and know where things are in space, one's sense of meaning will be impaired and learning will be affected. A visual problem, therefore, represents a distortion in the way one operates in visual space. When an individual demonstrates a distortion of his visual space world "he is also exhibiting manifestations of his particular visual problem" which will interfere with all aspects of learning and performance. Improve the individual's visual space world, rid him of his visual problem and his learning and his performance in such activities as reading will improve (Skeffington 1958c).

Chapter 19

The Visual Stress Response

Theoretically, visual stress is just one of many possible local stress reactions. Yet, its existence has very important ramifications. First, it is becoming evident that many visual problems, including some that were considered to be genetic, are basically stress induced. Second, the increasing visual demands of our society appear to be causing an increase in the frequency and severity of these stress-induced problems. And third, due to the importance of the visual process in doing, learning, understanding and remembering, stress-induced visual difficulties reduce the efficiency of those performance skills that are so necessary to cope effectively in our highly literate, visually oriented, technological culture.

Many of the factors involved in the visual stress response have already been touched upon in the discussions on general stress. They are worth reviewing from the vantage point of their specific relationship to vision and visual tasks.

Factors Influencing The Visual Stress Response

The intensity of the task, or more accurately the intensity with which a visual task is pursued, is one of the more important factors involved. Intensity relates to the degree of physiological arousal. The degree of arousal is related to the amount of attention given to the task. Attention is related to the mental effort invested. Mental effort is related to the requirements of the task, whether it involves loading (storing items in memory), retrieval of information, problem solving, categorizing or reasoning. As a general rule, the greater the intensity with which a visual task is approached, the greater will be the stress response.

The catatoxic (fight) approach is one in which the individual diligently engages in a task with maximum visual concentration, maintaining a strong desire for achievement, and attempting to get the work done in as short a period of time as possible. The syntoxic (flow) approach involves pursuing the task with the least amount of pressure or tension, flowing with it, not becoming upset if the work is somewhat difficult or if it takes a little longer than expected to complete, and feeling free to take breaks as needed without signs of guilt or remorse. The flight approach is one of avoiding the task or doing it with a minimum of involvement. It includes doing substitute things such as listening more in class in order to avoid reading at home, skimming when word-by-word reading is more appropriate, and daydreaming. It also encompasses being inefficient and just not caring. The fright approach consists of a paralysis of action, just staring at a task immobilized or being too apprehensive to proceed.

Impulsive individuals (Selye's "racehorses") who do their visual tasks while confined to restrictive quiet and tranquil environments tend to show an increase in the visual stress response as do reflective individuals (Selye's "turtles") who do their visual tasks in a hustle-and-bustle environment in the midst of many people and much noise and who are expected to work quickly. Analytical individuals who are logically and sequentially oriented and tend to invest more attentional energy into what they do, exhibit a greater stress response than those who are less analytical. Guilford (1967) calls this type of analytical approach to a task "convergent thinking." DeBono (1969,1968) refers to it as "vertical thinking." Field-dependent individuals demonstrate a greater visual stress response when doing a visual task than do those who are field-independent.

Detailed visual tasks such as technical reading or needlepoint result in a greater visual stress response than do less detailed visual tasks. Prolonged visual tasks are more stressful than shorter ones. Frequent and repetitive visual tasks are more stress producing than transient and more widely spaced ones. Reasoning tasks are more stressful than non-reasoning visual tasks since they entail more internalization and a reduction in external awareness.

Internalization, which involves a "detuning" of much of the environment, is more stressful than being wide open to one's surround. Therefore, intellectualizing, subvocalizing and internal verbalization when they are used in conjunction with a given visual task to maintain effective performance, will tend to increase the severity of the visual stress response. Similarly both input overload (when too much is going on or the input is too intense) and input underload (decreased input, social isolation and immobilization) which hinder integrated focal- global information processing tend to result in a greater stress response.

One's physical and emotional states, such as current illness, the effects of past illnesses, anxiety, fatigue, lack of exercise and poor nutrition, adversely affect the ability to adapt to visual stressors. They also lower one's "optimal stress level" and "optimal arousal level," thereby increasing one's vulnerability to the effects of stressor agents. Neurotic individuals tend to recover more slowly from visual stress while psychotic individuals tend to show a less regular or predictable visual stress response pattern.

The most important factor influencing the visual stress response is psychological state with attitudes, pre-sets, biases, motivations and perceptions affecting the stress response in various ways. Psychological acceptance of a visual task will make it less stressful than resenting it. Enjoying a visual task and finding it pleasureful will be even less stressful.

Input overload and underload are also reflected in one's attitude. Those in a situation of input underload may relate that they are bored with their work, that they find little variety in it and seldom look forward to going to work. They often feel that they are overqualified for their job and tend to demonstrate apathy and lethargy. Those in a situation of input overload usually express that they do not work well under pressure, that they get irritable over little things, that they do not sleep well and constantly feel tired and depressed. They tend to make more mistakes and are usually indecisive. They often take work home, feel tense, anxious and impatient. Finally, the perception that one is not in full control of a situation, the inability to predict what comes next and the feeling that one cannot escape from what one is doing are more stressful than the perception of controllability, predictability and escapability.

In terms of prior conditioning, once one responds in a certain way to a stimulus, it becomes easier to repeat that same reaction the next time one is exposed to a similar stimulus. In this way, responses to a given stressor become conditioned and generalized so that one learns literally to react to a low-intensity stimulus the same way one previously responded to a similar high-intensity stimulus.

The Visually Near-Centered Task

The Skeffington (1950a) visual stress concept can be concisely summarized by the statement that visual stress results from "...the biologically unacceptable, socially compulsive, visually near-centered task which provokes an avoidance reaction ... that becomes a drive to center nearer in visual space."

There are a number of important elements in this statement. The first has to do with the relationship between the "visually near-centered task" and the stress response. Too many clinicians have interpreted this to mean that near work is a major stress producer to the visual system, much more so than visual tasks done at intermediate and far distances. This conclusion seems to be supported by the traditional view that one has to accommodate more to see at near than at far. In the case of a prolonged near- vision task, the implication is also present that increased accommodative effort would have to be maintained for long periods of time. This, however, is not quite how Skeffington viewed it.

To Skeffington, accommodation in the pre-presbyope (those under 35 to 40 years of age) was never a primary problem since at this stage of life one should naturally have and be able to maintain sufficient accommodative ability for reasonable periods of time. Limited or inflexible accommodative ability in the pre-presbyope was seen as being the result of the visual stress response, not its cause.

Near-vision tasks were considered to be a major cause of visual stress not because of the nearness of the work or the amount of accommodation that is necessary but because of the containment factor and the type of intense meaning-getting that is demanded by the culture at that distance in space. If the same type of maintained attention is imposed at another area in space such as when using a VDT (video display terminal) at an intermediate range or at a far distance in the case of movie projectionists, then these other areas of space would become primary causes of visual stress instead.

The nearness of a visual task, therefore, is actually less of a problem than those factors that contribute to restraint of movement. As Skeffington (1953) put it:

> Not alone is the sheer nearness factor one of the reasons for the organism to make adaptation, but the fact that if he is to get meaning out of the thing that he is doing, he simply cannot exercise his freedoms to move. That is the reason why we use the words "socially compulsive." He is forced to stay on a task and thereby he absorbs, utilizes, the latitudes built in him....There is always a contraction in the freedoms to operate under any continued stress.

"The human organism builds significant degrees of latitudes in everything it does. These degrees of freedom contract or shrink when the individual is under stress....The socially compulsive visually near-centered task...is definitely the most stressful task that we know about because it makes its impact on the most elaborate organization demanded of the human being"--its search for meaning (Skeffington 1955b).

"The containment aspects of the bio-social world characteristically bring a constriction of the movement patterns involved, limit the ability to aggregate information and inevitably limit learning ability and internal communication." Again, "there is nothing harmful in looking at nearpoint. The problem is not the nearness. It is the containment factor....The environmental condition is a world of enclosures. Our bio-social tasks demand a restraint from moving; containment that says, 'Thou mayest not move'" (Skeffington 1961).

Over time, this concept expanded to include the stressfulness of problem solving, the importance of maintaining a proper operational balance between "figure" and "periphery" (Skeffington 1965a) and the recognition that a feeling of inadequacy can also add to the stress response (Skeffington 1966).

A second element in the Skeffington stress concept relates to the hypothesis that the visually near-centered task provokes "an avoidance reaction." Skeffington saw the culturally imposed near-centered task as being a demand that few could escape. He conceived that the "sheer genetic survival drive is to get away from any task that is too near, the word 'away' literally meaning a physical distance. The organism needs room to operate" (Skeffington 1957e). He noted that "there is a fundamental demand within the organism to escape from any too closely containing space. The surest way to arouse a frenzy of escape activities is to confine any person, young or old, so that movement is impossible" (Skeffington 1957f).

The third element in the Skeffington stress concept pertains to the visually near-centered induced avoidance reaction becoming transformed into "a drive to center nearer in space." This drive to center or localize closer in space is manifested visually as a tendency to overconverge (esophoria).

An overconvergence tendency has long been recognized clinically as a major cause of adverse symptoms and reduced performance ability. A mild overconvergence tendency usually produces more discomfort than a mild underconvergence tendency. Also, accommodative problems alone do not seem to cause the same degree of disturbance or discomfort as they do when they are accompanied by a tendency to overconverge.

The "Pull-In" Phenomenon

This drive to overconverge also appears to be related to what can be called the "pull-in" phenomenon. It has been a common observation that many people tend to come progressively closer to their work or bring their work closer to them as they get involved with a near-vision task. Some seem to do this as soon as they start a near task such as reading or writing. Others appear to do it after they have been working for a while. Still others demonstrate this response only intermittently in relation to the difficulty of the task at that point.

A study by Robert Phelan and Newell C. Kephart of Purdue University found that as the difficulty of a reading task increased, some children moved closer and closer to the page. Others moved sideways or occasionally backward. The pull-in-ers showed more restlessness than the other group when levels of difficulty were reached. Using a retinoscope, Drs. Phelan and Kephart noted an "against" motion of up to 3 diopters in this group indicating increased accommodation. They reasoned that those who moved closer to the task or brought their work closer tended "to put themselves into the task more wholeheartedly than the others" (Radler [1955] 1956).

Without doubt, some part of this reaction may be due to the tendency to overconverge or localize closer in space than where the task is positioned. In this particular instance, the "pull- in" phenomenon can be viewed as a psychological attempt to come to terms spatially or environmentally with the physiological drive to overconverge. In effect, it would be an ecological response.

However, there are many exceptions. Not all measurable overconvergers tend to bring their work closer or come closer to their work. Some pull away. Others just remain at their normal working distance. Some decided underconvergers exhibit the "pull- in" phenomenon. As Phelan and Kephart noted, some of the pull-iners demonstrate an "against" motion retinoscopically as they pull-in, usually coming closer to their work only in relation to the difficulty of the task. Some of the pull-in-ers who show an "against" motion demonstrate an overconvergence tendency when tested during and immediately after a demanding near-vision task. Many do not. This infers that the "pull-in" phenomenon may be related to a tendency to overaccommodate either with or without a corresponding tendency to overconverge.

It has been observed for some time that the stress response is associated with a tendency for the organism to constrict physiologically, functionally and perceptually. Kurt Goldstein (1939,1963) noted this in his work with severely injured subjects. Virginia I. Shipman (1954) stated that as a result of this stress-induced perceptual constriction, the individual begins to "observe less, see less, remember less, learn less and become generally less efficient." Constriction always occurs for the purpose of survival. It is a basic biological reaction to protect the organism and is similar to the turtle contracting into its shell or the porcupine rolling itself into a ball. In this respect the tendency to overconverge and overaccommodate can also be viewed as part of this organismic "constriction" reaction.

More important is the role that figure-ground organization plays in this "constriction" response. As was pointed out earlier, it is a general rule that the more focal (central) an individual becomes, whether necessitated by the task or by one's attitude to the task, the greater will be the tendency to tune-out extraneous global (peripheral) information. In this light, the "pull-in" phenomenon can be seen as an organismic drive to foster focal attention while repressing, reducing or subduing the peripheral retinal input. It is an attempt to close out possible external distractions in order to become more totally involved with the situation at hand.

The Skeffington Stress Model

The mechanics of the basic Skeffington stress model were never presented in detailed physiological terms. Its operation was originally explained grossly as a four-part action on the part of the visual system as it responded to near-visual stress. The first part starts with the assumption that theoretically, at any point in space where one sees clearly and singly, one should be using one unit pattern of accommodation (1A) and one unit pattern of convergence (1C).

The second element in the theory holds that visually near-centered stress sets up a visceral dominance which results in the activation of two unit patterns of accommodation (2A) to one unit pattern of convergence (1C).

The third element hypothesizes that since both accommodation and convergence must operate in unison, the organism will tend to produce two unit patterns of convergence (2C) to match the increased accommodative response.

The final element in the model presupposes that the increased convergence response will tend to jeopardize seeing clearly, singly and comfortably unless it is blocked so that the organism can somehow maintain the distorted relationship of two units of accommodation (2A) to one unit of convergence (1C). To restate this in Skeffington's (1957a) own words:

> Consider the movement patterns of identification and centering at near. Label them "A" and "C." Put the organism under stress at nearpoint. Designate the greater change in movement pattern of identification by "A" and "A." Remember, there is no loss of acuity. In fact, the change is made to preserve the ability to maintain "contour pattern discrimination" (acuity). The related pattern of centering should (then) change in like degree. However, if it did so, it would move to the less effective end of its range or even to a point where alignment would change and diplopia would result. This movement must be checked (inhibited).

Active inhibition is tension. Tension results in constriction and decreased performance. "The only way that this cycle of difficulty can be prevented from continuing would be to find some way to allow the completion of the drive to center nearer" (Skeffington 1964). Theoretically, if there are no other complications, a plus (convex) lens prescribed for near use should allow this to happen. It permits the individual to converge or center for a closer point in space than where he is accommodating, without the need to inhibit actively convergence, thus allowing him to have single, clear and comfortable vision without tension.

This is only a small part of Skeffington's contribution to the visual stress concept yet these few ideas alone revolutionized clinical visual care, spurring the growth of what has become known as behavioral optometry. The concept was a radical departure from the conventional wisdom of the day which treated the eye as a camera and visual problems as optical defects that one either inherited, developed due to disease or age, or suddenly acquired for no discernible reason. The Skeffington model postulated that:

1. the visual process was responsive to the interaction of the individual with the environment,
2. a change in visual status occurred for very specific reasons,
3. the most frequent reason for this change was the containment factors involved in the increased prevalence of culturally imposed near vision tasks,
4. these containment aspects of near vision tasks initially act through the viscerally controlled identification ("what is it") process by stimulating its accommodative component,
5. the most potent and potentially negative effects on the visual process occur not with this increased accommodative response but with the secondary stimulation of the alignment and convergence components of the centering ("where is it") process to foster a localization that is closer in space.

Skeffington hypothesized that the reason this overconvergence tendency was the major negative factor in the visual stress response was that it could only be restrained by active cortical inhibition and not by any automatic, lower-order, self-regulating homeostatic mechanisms.

A Physiological Reconciliation

As was mentioned in chapter one, I became personally involved in the intricacies of this Skeffington visual stress model when I tried to reconcile its elements with existing physiological knowledge. To do this meant drawing on the thoughts of Claude Bernard, Walter B. Cannon, C. Judson Herrick and, of course, Hans Selye. Their work stressed the action of the autonomic nervous system and indicated that the initial response to stress was a sympathetic nervous system reaction. In the visual process, this would be represented by a decrease in accommodation and an enlargement of the pupil. The Skeffington model did not emphasize the actions of the autonomic nervous system and started with an increase in accommodation with no mention being made of pupillary variations during stress. I questioned a number of leading behavioral optometric theoreticians, including Dr. Skeffington, about this particular discrepancy. The responses were all quite similar in nature. No one agreed or disagreed. On the part of many, there was an apparent lack of in-depth knowledge of the physiological processes related to stress. What was disappointing was that there appeared to be little interest in pursuing the issue. The "black box," stimulus-response approach was sufficiently productive at that point in time so that there was little desire to delve into more physiological areas. Only one person had a working acquaintance with the issues involved-- Darell Boyd Harmon.

Harmon was a noted expert in such diverse fields as pediatrics, kinesiology, architecture, lighting, engineering, pedagogy, penology, human performance and vision. Though he was not an optometrist, his scope of knowledge was such that it left a strong imprint on behavioral optometry. At the time in question, Harmon was doing research on the effects on vision of lenses, posture and shifts in working distances. His major tools were retinoscopy and electromyography. He reported the occurrence of an "against" motion with the retinoscope (indicating increased accommodation) and a constriction of the pupil in two areas of space as the subject observed a test object. The first, a weak reaction, occurred as the fixation target entered inside the limits of the body's extension (the full reach of the arms) where touch, during the developmental years, guided and reinforced the visual act. The second and stronger reaction was noted when the test object reached that area in space where the individual habitually held his reading material while striving for maximum meaning. Both reactions indicated that not only could stress responses be conditioned but that they would relate to the areas in space in which they occurred. These observations later became the basis for the development by Robert Kraskin (1965) of the clinical technique known as "stress-point retinoscopy."

After hearing Harmon present the results of his research, I discussed the subject with him and mentioned that according to stress theory a rapid sympathetic reaction should precede the parasympathetic response. He replied that this was exactly what he observed. Every time he saw an "against" motion and a pupillary constriction, it was preceded by a fleeting "with" motion and pupillary dilation. This confirmed that the initial visual stress action was a sympathetic one and that in the unconditioned response, this sympathetic response generally occurred for only a limited period of time before being followed by the long-observed parasympathetic reaction.

I then proceeded to incorporate these ideas more formally as part of the fifteen-part series, A Modern Philosophy of Vision, that was published in 1959 (Forrest 1959-60a). Elements of this physiological approach have since been supported in papers by Gawron (1983) and Birnbaum (1984b). A further, more detailed probing of some of the physiological processes at play in the visual stress response will be the subject of the next chapter.

Chapter 20

Physiological Aspects of Visual Stress

Models, whether physiological or behavioral, are always less than ideal. Due to obvious limitations in knowledge, they either turn out to be completely wrong at the worst or partially right at the best. This does not prevent them from being useful either as pegs to hang observations on, as explanatory frameworks for communication with others or as a springboard for further intellectual growth. Models are always supported by certain key relationships which help explain them. It is some of the underlying key relationships relating to the visual stress concept that we will explore at this point.

There are two major functional components to the nervous system: the somatic (skeletal) and the autonomic (visceral). The somatic aspect of the nervous system controls the skeletal musculature of the body while the autonomic aspect acts predominantly on smooth internal muscles. Stimulation of the somatic system results in a rapid response as compared to the slower action of the autonomic system. Stimulation of efferent motor neurons of the somatic system results in excitation only. There are no inhibitory nerves in the skeletal system. Inhibition comes only from the central mechanisms of the brain which act on the motor neurons directly and not on the muscle fibers.

In the autonomic system, there are two subdivisions, the sympathetic and the parasympathetic, which act as antagonists operating wherever possible to maintain an innervational balance (see Figure 23). If one stimulates, the other inhibits. If one is in ascendance, the other is in descendance. Unless a response is conditioned, there is always an attempt made to restore equilibrium. In the autonomic (visceral) nervous system the ability to inhibit is built-in as part of the action of these dual antagonists. Although the system operates to a great degree automatically, it is highly influenced by central cortical control which, in turn, is influenced by attitudes and emotions (Gillian 1954).

Vergence eye movements (convergence and divergence) are functions of the somatic system. They result from innervation to the extraocular muscles surrounding the eye. Convergence is a result of third cranial (oculomotor) nerve stimulation acting on the internal or medial recti muscles. Divergence is a result of sixth cranial (abducens) nerve innervation to the external or lateral recti muscles.

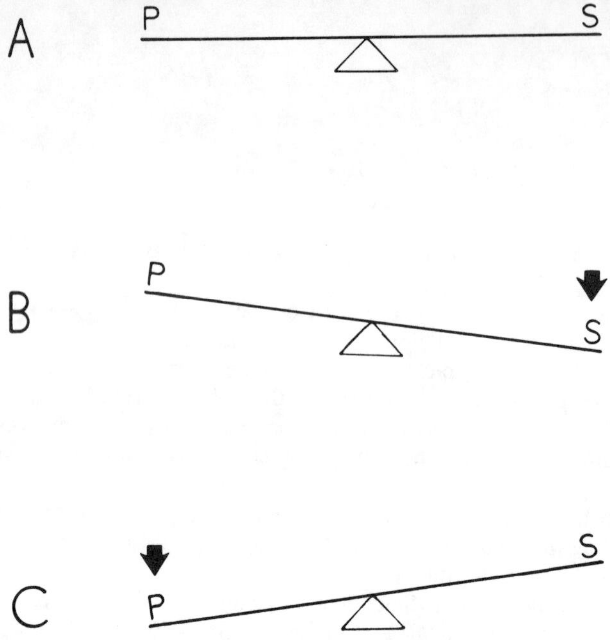

Fig. 23. The equilibrating action of the autonomic nervous system

Accommodation is primarily a function of the autonomic nervous system and results from innervation to the ciliary muscle controlling the crystalline lens of the eye.

Parasympathetic stimulation to the eye results in: (1) an activation of accommodation (an increase in the focusing power of the lens) through the contraction of the circular fibers in the ciliary body, (2) pupillary constriction, (3) a narrowing of the lid opening due to a combined drooping of the upper lid and elevation of the lower lid, and (4) a retraction of the eyeball.

Sympathetic stimulation to the eye results in: (1) a relaxation of accommodation (a decrease in the focusing power of the lens) by a contraction of the radial fibers in the ciliary body, (2) pupillary dilation, (3) a widening of the palpebral (lid) aperture, and (4) a slight bulging of the eye.

There is also a definite link between the action of accommodation and convergence. Each affects the other. Parasympathetic activation of accommodation results in stimulation of the third (oculomotor) nerve causing the eyes to converge. Oculomotor stimulation of the medial recti muscles of the eye to converge results in an increase in parasympathetic stimulation to accommodation. It also causes a narrowing of the palpebral aperture, retraction of the eyeball and constriction of the pupil. That degree of convergence that is induced by accommodation is called accommodative-convergence. That degree of accommodation that is induced by convergence is known as convergence-accommodation.

Little is known of the actual cause of these interactions though there has been much speculation on the subject. One hypothesis is that in the tectal or diencephalic region of the brain, there appears to be a connection between the Edinger-Westphal nucleus supplying the ciliary ganglion and the nucleus of the medial recti in the third nerve complex. A second view holds that since the parasympathetic nerves to the eye travel in close physical proximity to the third (oculomotor) nerve, this could result in each influencing the other. A third view notes that the post- ganglionic parasympathetic fibers are mediated by acetylcholine which is also the transmitter used by the somatic (skeletal) system in its neuron-neuron and neuron-muscle connections. It is therefore hypothesized that somatic neurons adjacent to parasympathetic neurons could be responsive to the secreted acetylcholine.

The Intermediate Dark Focus Concept

For many years it was taken for granted that an emmetropic eye looking at distance is completely relaxed. As the eye focuses to see objects closer to the individual it has to use progressively more effort; zero diopters to see at far distance and 2.50 diopters to focus at the average adult near-reading distance of 16 inches. This has led many investigators to conclude that parasympathetic action is the primary visual mechanism for accommodation. An increase in parasympathetic innervation was hypothesized as being needed to see at near with only a decreased innervation being necessary to see further away. Sympathetic action was assumed to be negligible in the act of accommodation, perhaps simply supplying a tonal background for the operation of the parasympathetic system (Davson 1976). This view ironically was contrary to the way the sympathetic and parasympathetic systems were known to interact with each other in other parts of the body. In recent years, a newer and different perspective has come about with the discovery of the intermediate dark focus of accommodation.

Dark focus is another name for the resting state of accommodation or that position in space where no accommodative effort is in play. Studies using a laser optometer have demonstrated that in darkness the resting state of accommodation is at an intermediate distance rather than at infinity where it was always thought to be. Leibowitz and Owens (1975) originally found the mean dark focus to be 1.7 diopters (59 centimeters or 23.2 inches) with a standard deviation of 0.72 diopter and a range of approximately 4 diopters. They later modified this slightly, reporting that the dark focus averaged 1.52 diopters with a normal distribution over a zero to 4-diopter range (Leibowitz & Owens 1978). Westheimer and Blair (1973) noted in laboratory animals that in darkness, in sleep and under anesthesia the resting state of accommodation shifted to the intermediate dark focus. At death, however, accommodation shifted outward to zero diopters. This led them to differentiate between the condition where no accommodation is in play and true anatomical rest when there is no innervation to the focusing muscles at all.

Though the resting point of accommodation exhibits considerable variation from person to person (Leibowitz & Owens 1978) it is usually quite stable for a given individual over time (Mershon & Amerson 1980, Miller 1978, Owens & Higgins 1983). The fact that the dark focus is at an intermediate position in space (see Figure 24B) rather than at distance (see Figure 24A) indicates that the sympathetic and parasympathetic systems act in the eye as dual antagonists (or as a dual innervation process) in a similar manner to the way they interact in other body systems, with the dark focus representing a passive state of tonic balance of sympathetic parasympathetic action (Gawron 1983). As long as the resting state of accommodation was thought to be at infinity, there was little reason to suspect the sympathetic nervous system of being a major influence in the act of accommodation.

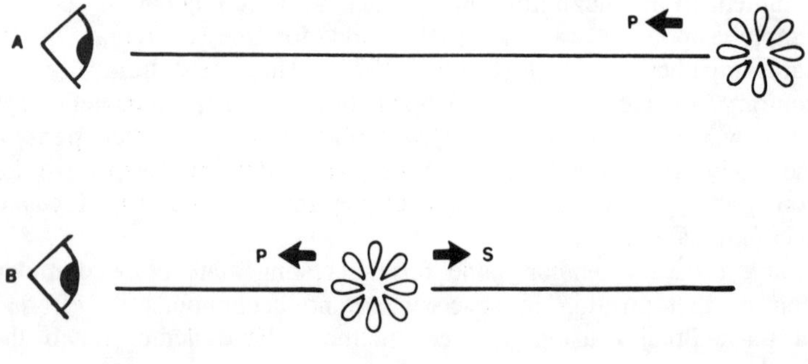

Fig. 24. The intermediate dark focus of *accommodation*

Besides the fact that the resting state of accommodation (dark focus) is located at an intermediate point in space, there also appears to be a bias for accommodation to posture at the dark focus regardless of where one is fixating (Owens 1979). This bias is greatest in darkness or when the accommodative stimulus is degraded such as in the presence of low illumination, reduced contrast, reduced discrimination ability, an absence of visible contours or when engaging in a concurrent visual task (Leibowitz & Owens 1975, Johnson 1976). This bias is reflected in what has come to be called the lead and lag of accommodation. If accommodation is less than expected or postures further away when one focuses on a near object it is said to "lag." If accommodation is greater than expected or postures nearer in space when one focuses on a distant object, it is said to"lead." This tendency for accommodation to lag and to lead increases the further the object being looked at is from the dark focus. Visual acuity and accommodative accuracy are found to be at their greatest when the object of regard is placed exactly at the dark focus (Johnson 1976). This indicates that the lag and lead of accommodation are a functional compromise that places accommodation between the dark focus and the actual location of the stimulus (Owens & Higgins 1983). This also offers an explanation for the existence of the anomalous myopias (night myopia, empty field myopia and instrument myopia). These are instances where one measures more myopia at night, when looking into empty space for long periods as pilots and astronauts often do, or when looking into optical instruments such as microscopes. These myopias are now being considered to represent the passive return of accommodation toward the resting state (dark focus) when the stimulus is degraded. They also occur when accommodation is ineffective in changing the retinal image as usually happens when looking into optical instruments with small exit pupils and an extended depth of field so that there is less need to accommodate over a wide range of optical distances (Leibowitz & Owens 1975).

The "lead" of accommodation may be a new concept but a "lag" of accommodation has been noted since dynamic retinoscopy first came into use. Dynamic retinoscopy is a technique that measures the accommodative response while a subject attends to a stimulus such as reading material, an object or a light placed at the plane of the retinoscope. Under most normal, nonstressful conditions a "with" motion is observed, indicating that the individual is focusing beyond where the object he is regarding is located. It is this disparity that has been traditionally known as the "lag of accommodation" (Cross 1911).

In chapter 9 it was mentioned that in the technique known as book retinoscopy, a variant of dynamic retinoscopy, it was common to find an "against" motion reflex during periods of intense problem solving. Those who consider that this is an artifact of the test have used the more rigidly controlled MEM technique (Haynes 1960; Greenspan 1974; Rouse, Hutter & Shiftlett 1984; Rouse, London & Allen 1982) which has more consistently revealed a lag of accommodation of between +0.50 and +0.75 diopters. Rouse et al. (1984) in a recent study, found the mean lag to be approximately +0.33 diopters +/-0.35 diopters, indicating that the average expected range of the lag is between Plano and +0.75 diopters.

One question that arises is whether this lag represents an active sympathetic nervous system response or not. Some consider that it does while others feel that it just represents a tendency to posture accommodation closer to the dark focus position (Gawron 1983). A more important question is whether the lag of accommodation is something that has to be overcome with effort in order to see clearly and meaningfully at near point. Although some seem to think that this is so, it should be noted that the measurement of the lag ("with" motion) during the MEM procedure is found while the subject is reading clearly and apparently meaningfully on the target that is at the plane of the retinoscope. This alone would indicate that being focused for a point beyond the plane of regard may be a useful if not needed correlate of efficient meaning-getting rather than being a disparity that has to be overcome with effort.

Mental states (emotions, attitudes, thinking, imaging, etc.) are also highly related to where accommodation tends to posture. Westheimer (1957) found that he could get increases in accommodation of up to 1 diopter for periods as long as five minutes after insulting his subjects. Malstrom and Randle (1976) were able to accelerate or retard the inevitable empty field myopia drift to the resting state value of 1 diopter by having their subjects use visual imagery and "think near" or "think far."

Leibowitz (1977) observed that anxiety and anger usually result in an increase in accommodation. Costello (1974) found that looking at slides of automobile accidents tended to result in increased accommodation while relaxation procedures fostered a decrease in accommodation. Olmstead (1944), however, noted that individuals subjected to sudden stressors such as loud sounds or electric shock demonstrated a momentary reduction or outward shift of accommodation. Gellhorn (1953), after reviewing the literature, accounted for some of the divergent findings by noting that an angry response to a stimulus results in a completely different nervous system reaction than does an anxious response. Fear, he concluded, results in sympathetic activation while anger and anxiety appear to result in a predominantly parasympathetic reaction.

The intermediate dark focus concept has also provided a new rationale for lens prescribing. In the traditional view, the resting point of accommodation is at far point or infinity. In order to focus at 16 inches, the individual has to use 2.50 diopters of accommodation. Most practitioners have noted that a low power plus lens (+0.50 to +0.75 diopters) is usually all that is needed to reduce fatigue, restore comfort, increase efficiency and minimize autonomic nervous system reactivity when doing nearpoint visual tasks (Pierce 1966-67). A +0.50 diopter lens is 20 percent of the 2.50 diopters needed to focus at near and only 10 percent of the 5.00 diopters of accommodative amplitude that is usually available to the pre-presbyope. Because the +0.50 lens is such a small part of the accommodative effort being expended, many have found it hard to understand why it tends to be so effective. Reducing one-third to one-half the effort has always appeared to be a more plausible and clinically accepted method of prescribing for ocular muscle tension problems.

The dark focus concept places the resting point of accommodation not at infinity (zero diopters) but at approximately 1.5 diopters. This reduces the total accommodative effort that is thought to be needed to focus at near. If one thinks of effort alone, this offers a more reasonable hypothesis for the effectiveness of prescribing low-power plus lenses. Weisz (1980) relates the most effective lens power prescription to the lag of accommodation as it, in turn, relates to the dark focus. She considers that low-power plus lenses allow the accommodative response to assume a position closer to the resting state which would tend to reduce autonomic nervous system activation (Pierce 1967) and perhaps allow more detail to be extracted from the stimulus (Johnson 1976).

The Intermediate Dark Vergence Concept

Like the accommodative system, the traditional view of the vergence system considered the starting point of convergence to be at distance (optical infinity) where the eyes were parallel. Theoretically, this physiological position of rest involves some degree of vergence activity, called tonic convergence, to bring the eyes from the anatomical (slightly outward) position of rest to parallelism. Maddox (1893) felt that all convergent eye movements inward from the physiological position of rest at infinity involved effort while divergence was simply a passive process of relaxation, a reduction in muscular effort.

The more recent view of the action of the vergence system supports a dual antagonist or opponent-process hypothesis that postulates that a balance of innervation to the lateral and medial recti muscles causes the eyes to rest at an intermediate position in space when stimulation to the visual system is absent (Owens & Leibowitz 1983). This has become known as the dark vergence position. Margach (1983) suggests the use of the term dark locus as a more appropriate name for the intermediate position of both accommodation and convergence.

Like the action of the autonomic nervous system in relation to the dark focus of accommodation, convergence and divergence from the intermediate resting position are considered to be active processes. This infers that both near and far activities require effort that could result in fatigue or discomfort.

The resting state of convergence has also been found to be biased in the direction of a previous vergence response to prism. This has been called vergence adaptation. It was determined that looking through a prism that either stimulates or inhibits convergence results in a change in phoria measurements taken after the prism is removed (Schor 1983). Schor (Schor, Johnson & Post 1984), investigating the after effects of prolonged monocular accommodation responses in relation to the dark focus of accommodation also found that the accommodative system adapts similarly to the way the vergence system adapts, shifting its position after prolonged accommodation for a near point in space. Ebenholtz (1983) noted that these adaptive aftereffects are apparent after only eight minutes of accommodation.

These studies indicate that prolonged and concentrated near work can trigger a shift in the positions of both the dark focus and the dark vergence in the direction of the near-stress demand. As the dark focus and dark vergence positions shift inward, there is a resulting reduction in the accommodative and convergence effort needed for doing the near task. This process may be another example of a system moving in the direction of the stress in order to reduce or minimize the stress.

The accommodative system has been found to adapt more slowly than the vergence system and is more highly variable both between and within subjects. Schor (Schor, Johnson & Post 1984) concludes that these adaptive effects may, under certain circumstances such as fatigue and reduced visibility, interfere with distance activities by increasing the amount of night myopia. The implication is that adaptation to prolonged close work can cause the dark focus to move inward, resulting in a greater degree of accommodative "lead" to occur when looking at distance due to the increased disparity between the dark focus and the distant object of regard. This offers a possible explanation for the cause of what has been called "accommodative spasm," "pseudomyopia," or "temporary myopia."

Wolf (1984) found that the magnitude of change in both dark focus and dark vergence is dependent on the individual's initial resting position. Those with a dark focus and dark vergence that is initially set further away than average are more affected by close work and tend to experience more intense symptoms of visual fatigue since more effort has to be expended to focus and fuse at near. This results in still greater adaptive shifts of accommodation and convergence inward. Wolf noticed a lack of correlation between changes in dark focus and dark vergence indicating that they respond independently. She also noted that fatigue was more highly related to changes in dark vergence than to changes in dark focus. This observation lends some support to the Skeffington model in which an increase in convergence due to the effects of near work is hypothesized as being the more potent factor in the visual stress syndrome.

Chapter 21

A Behavioral Model of Visual Stress

At this point I would like to outline the essential aspects of a personal behavioral model of visual stress that is based on the functional philosophy of A. M. Skeffington. It is a framework that I have used and felt comfortable with for many years. Admittedly it is an "as if" model, a metaphor for what appears to be happening. Even with this limitation, it has served productively for organizing many diverse clinical observations into an effective structure.

For ease in communication, a few of the key actions of the visual process will be shown in schematic form. The basic diagram (see Figure 25) highlights the visual process as a dual effector, visceral-skeletal system. The vertical line represents the plane of regard at nearpoint, the area in space where most visual stress responses are first manifested. The closed rectangles labeled (A) and (C) represent the mechanisms of accommodation and convergence as they operate within the processes of identification and centering which are represented by the two larger, open-ended rectangles.

To restate what was mentioned in chapter 18, "centering," to Skeffington (1950b), "is the action of the whole organism toward whatever item in the environment holds greatest import for the organism at the moment." It is "the selection of some area in space for attention and meaning" (Skeffington 1958a) so that one can "know exactly and precisely where things are in space" (Skeffington 1957d). The goal of centering is to provide the knowledge of where something is, and this can be attained by central or peripheral means. It is not dependent solely on accurate eye alignment. Convergence, on the other hand, is the mechanical and anatomical alignment mechanism necessary for the visual axis of each eye to cross at a particular point in space when such alignment is required.

"Identification is the ability to discriminate and define, to obtain an adequate contrast boundary of the item under inspection, to form a visual datum that will be satisfactory to the organism" (Skeffington 1950b). It is the ability "to aggregate experiences from single, organized units of experience...by means of combining the enormous amount of experiences that must precede it" (Skeffington 1957). Identification provides the knowledge of what something is and, like the centering process, this can be attained by central or peripheral means and not necessarily solely by achieving a good focus on the retina. Accommodation, on the other hand, is simply the mechanical change of focus necessary in order to produce the best possible light distribution on the retina when it is required.

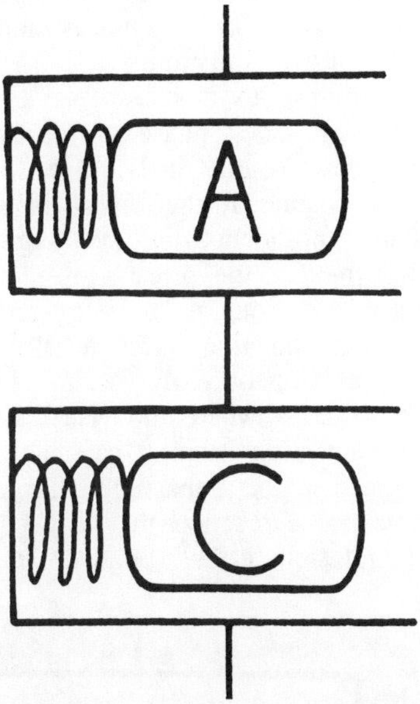

Fig. 25. The visual process as a dual effector system

The "springs" in the diagram represent the homeostatic mechanisms (the degrees of freedom or the buffering effects) of accommodation and convergence as they operate within the processes of identification and centering. The size and location of the large, open-ended rectangles represent the flexibility and the stability of the processes of identification and centering as they function on either side of the plane of regard for maximum localization and meaning-getting. This positioning also emphasizes the fact that these processes are spatially oriented, geared toward aiding the organism to function in space. The placement of the two open-ended rectangles with respect to each other illustrates the close relationship of identification to centering and accommodation to convergence, a relationship which, though genetically programmed, is predominantly learned.

Figure 25 depicts the key elements of the visual process in a theoretical state of balance. This state of equilibrium is one which may never be found to exist in a living, dynamic organism. It is being used here simply to symbolize a hypothetical starting point for the various actions that are seen to occur in the visual stress response. In a broader sense, it represents the physiological base from which the organism acts, reacts and adapts to environmental forces, internal physiological changes and to the attitudes, motives, needs, outlooks, prejudices and biases that one brings to a particular visual situation.

The initial visual response to a stressor is a sympathetic nervous system reaction. The shock phase of the Alarm Reaction of the visual stress response results in a drive for the mechanism of accommodation to posture further away in space than where one is both localizing and identifying (see Figure 26). It should be noted that in this model, the process of identification is considered to remain stable as long as meaning-getting is still being achieved at that spatial position. The usually short-term sympathetic reaction to a stressor only occurs in its purest and simplest form in those cases where the visual system responds in an unconditioned fashion. Conditioning can cause the sympathetic response to be augmented or bypassed. This initial momentary outward movement is not a reaction that is peculiar to the visual system alone. Kurt Goldstein (1939) pointed out that in any nor-

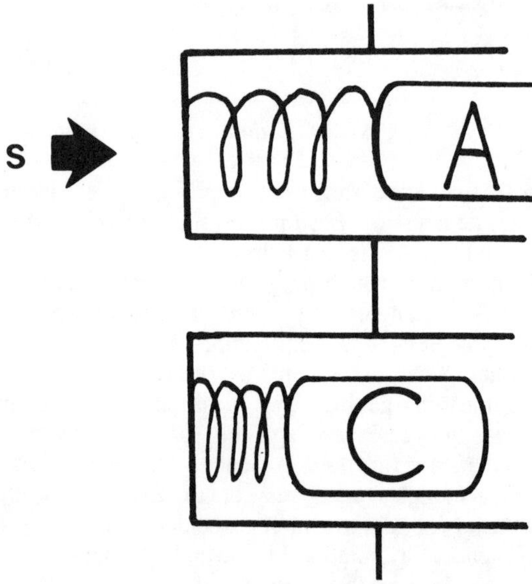

Fig. 26. The initial visual stress response

mal "grasp" situation where one pulls in, an outward spatial movement always precedes the inward movement. All stimuli (and stressors are stimuli) are brought within the grasp of the organism by first actively reaching out through extensor mechanisms and then by a constriction inward toward the organism via the flexors.

The tendency to posture accommodation outward in space in relation to both the identification process of which it is a part, and to the centering process creates an internal dissonance that the organism attempts to resolve in some way. The point is being stressed that it is possible to accommodate for a point in space other than where identification is postured for meaning. This often occurs when one attends to a peripheral stimulus or is involved in internal chatter or fantasizing. In any event, there is still a drive or bias to reconcile the discrepancy between the action of the mechanism and that of the process. The reason for this is that meaning-getting is at its maximum when accommodation and identification are coincident or at least in better balance. Similarly, though it is possible to function with a dissonance between the processes of identification and centering or between the mechanisms of accommodation and convergence, there is still an internal bias that favors a unity in their interaction. They function together best with the least amount of dissonance and, therefore, there is an organismic advantage toward minimizing all forms of imbalance whenever possible.

The dissonance which is a result of the initial sympathetic stress reaction triggers an attempt at reconciling the disequilibrium that is created. There are many possible avenues by which the organism can establish some form of functional compromise in order to cope with the situation. Of these, four adaptive styles stand out because they are more frequently observed clinically.

The first two are very similar in their basic goal. They both attempt to restore the mechanism of accommodation so that it is more in balance at the plane of regard and more coincident with the action of the process of identification (see Figure 27). The processes of identification and centering tend to maintain their positioning as long as meaning-getting and localization are being achieved.

These first two major adaptive styles are mainly differentiated by the method used to restore accommodation to the plane of regard. The first coping style uses the localization system to counter the effect of the sympathetic stimulation. It does this by using convergence-accommodation to restore accommodation to the plane of regard and maintain it there against any stimulus to posture accommodation further away in space. The second coping

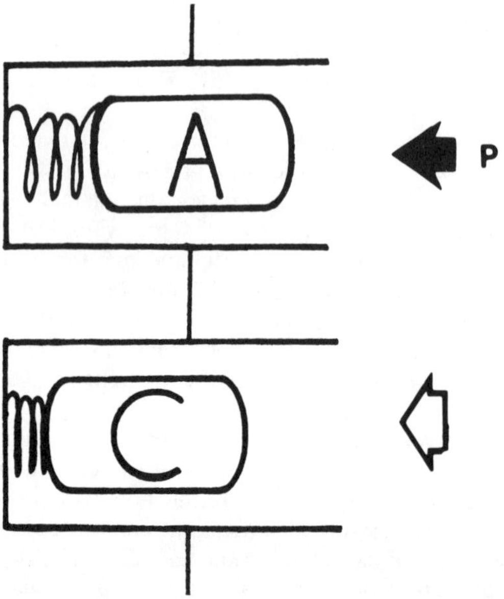

Fig. 27. The initial visual adaptive response

style relies on parasympathetic activation to compensate for the imbalance created by the increased sympathetic action. In effect, it utilizes accommodative-convergence to restore accommodation to the plane of regard and maintain it there. We will be discussing case types more fully later in this chapter but for now it will suffice to consider the first coping style which relies on convergence-accommodation to achieve its goal as being representative of the B1 type case and the accommodative- convergence approach as being representative of the B2 type case.

Figure 27 illustrates the two methods that can be used to restore accommodation to the plane of regard. The dark arrow represents the parasympathetic approach (accommodative-convergence) and the light arrow represents the predominant use of convergence- accommodation to resolve the problem. The shift in the position of convergence and the contraction of its spring illustrates the point that regardless of which adaptive mode is used, in both instances convergence will tend to posture closer in space than where accommodation is situated. So we see that an attempt to modulate the initial sympathetic stress response induces a secondary overconvergence tendency.

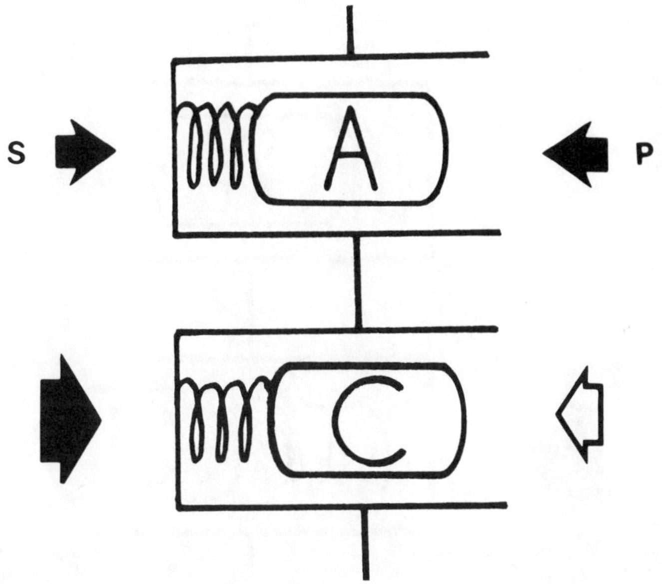

Fig. 28. Reducing the secondary overconvergene response

Subtle accommodative instability can be controlled by sympathetic-parasympathetic interaction (see Figure 28). Reduction of the over-convergence tendency, however, primarily occurs through central cortical inhibition (heavy black arrow). Any attempt to utilize increased parasympathetic stimulation to accomplish this through negative accommodative- convergence will tend to drive accommodation outward in space. This is counterproductive since it is actually going against the direction of the adaptive process.

There are two other less common visual adaptive styles. These styles also have their origin as a response to the initial sympathetic nervous system activation. The third coping style maintains a stability of accommodation through sympathetic- parasympathetic interaction but cannot maintain an active inhibition of the induced overconvergence tendency. In this instance, convergence collapses resulting in a receded nearpoint of convergence, frequent suppressions and fusional instability (see Figure 29). This is representative of the C type case.

The fourth visual coping style is a still less common occurrence in which the sympathetic response is maintained and where there is little or no attempt on the part of the parasympathetic sys-

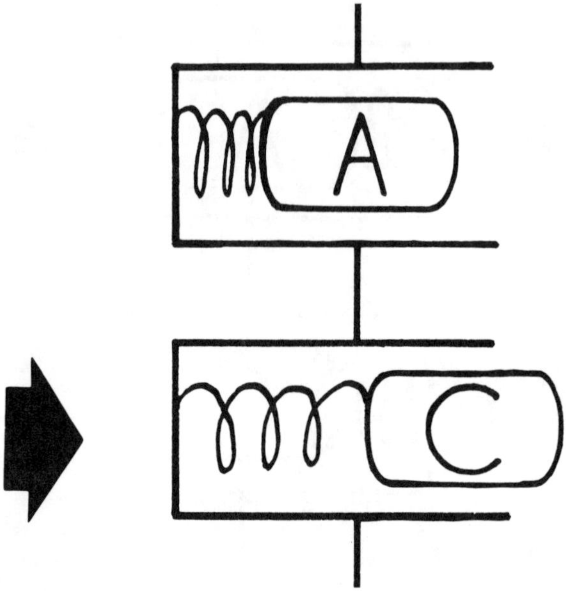

Fig. 29. The collapsed convergence response

tem to restore accommodative equilibrium. Convergence, in turn, recedes in an attempt to restore some balance with accommodation. Since both functions are detuned, away from the plane of regard, acuity will be degraded and convergence and accommodative skills will be inefficient and often inconsistent. Since both accommodation and convergence are localized in similar areas of space, there is usually no significant refractive anomaly noted (see Figure 30). At one time, I considered this approach to be "non-adaptive" (Forrest 1960b). Actually, it is quite adaptive although it is atypical. It appears to be brought on by a psychological detachment, a pulling-back from reality as a possible way of reducing anxiety. Often all that is needed to restore a more efficient adaptive approach is to "shake" the system, so to speak, through the use of low-plus lenses which posture accommodation further out in space, thus re-establishing the need to adapt in a more appropriate manner, similar to resetting a thermostat. This coping style can be considered to be a physiological rationale for what John Streff (1962,1969a,1969b; Appel & Streff 1963) - has called the "Non-Malingering Syndrome" or "Juvenile Bilateral Functional Amblyopia."

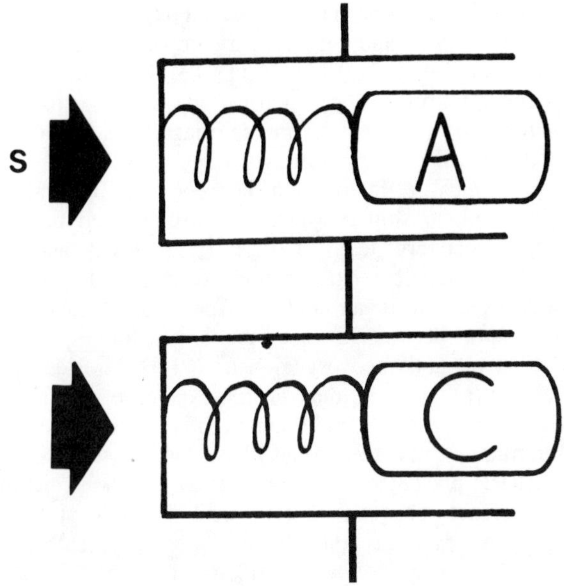

Fig. 30. An atypical adaptive response

The B1-B2 Case Types

Skeffington (1931) first noticed that clinical findings that varied from what were expected tended to cluster into groups or syndromes. He isolated three main syndrome clusters and called them diagnostic case types A, B and C. The A case type was considered as being brought about by some systemic toxic interference. The B case type, which was the most common, was thought to be representative of accommodative insufficiency. The C case type was considered to be representative of convergence insufficiency.

In time, Skeffington (1950d) began to consider the B case type as a syndrome representing adaptation to stress and being found in two forms, B1 and B2. "The altered behavior pattern which is a 'B1 case type' is a characteristic response of the organism to something in the environment. The 'B1' type...(is)...an individual who has made the characteristic response to the socially compulsive, visually near-centered task,...a drive to center nearer in space than the localization of identification." "These are all tasks that have social significance but no survival significance" (Skeffington 1950e).

What was called the B2 case type was considered to be a distortion of the B-type syndrome. It was thought that the B2 case started out as a B1 and, if treated properly, would revert to being a B1 case type. Skeffington (1958a) hypothesized that the B2 case stemmed from a very intense impact. "It seems to register the adaptation to a stress of larger magnitude, rather abruptly applied and continued long through time."

The major clinical differentiation between the B1 and B2 case types was the relationship between the base-out and base-in prism to break and recovery tests at near (tests #16B and #17B in the Skeffington analytical examination sequence). In the B1 case, the base-out prism to break and recovery test at near (#16B) is found to be low. In the B2 case, the lower finding is the base-in prism to break and recovery at near (#17B). Besides the statement that the B2 type is due to more intense stress, the relation between the base-out and base-in prism tests at near is the only differentiation made between these two case types.

In the model presented in this chapter, it was hypothesized that the difference between the B1 and B2 case types is due to the adaptive response style to the initial sympathetic activation. The B1 case type was considered to represent a convergence- accommodation approach to restore accommodation to a proper balance at the plane of regard. The B2 case type was noted to represent a parasympathetic accommodative-convergence approach to the situation. Both of these routes appear to be viable alternatives with different individuals taking one or the other of these two paths. This model proposes that regardless of which mode takes priority, the end goal is the restoration of accommodation to the plane of regard so that it is coincident with the processes of identification and centering.

Many years ago optometrists Nathan Flax and Daniel Woolf (1964) proposed a hypothesis which saw the B1 case type as attempting to restore centering out to the plane of identification and the B2 case as attempting to bring the plane of identification in toward the plane of centering. They considered that the B1 case type would rather know where something was rather than what it was and would therefore prefer to localize than identify. The B2 case type was seen as preferring to know what something was than where it was, to identify rather than to localize.

More recently Flax (1984) has related the processes of identification and centering and the B1 and B2 case types with Trevarthen's (1968) use of the terms "focal" and "ambient vision." The ambient system is viewed as a "where is it" system (similar to centering) and the focal system as a "what is it" system (similar to identification). Flax (1984) concludes that in city life there is more emphasis on time, how long it takes to go from one point to another, than on how far something is in terms of distance. In this way, since "what is it" decisions are apparently more important to urban life, one will tend to see more B2 cases in cities than in rural areas where hunting and farming ("where is it" functions) are more predominant.

The Trevarthen concept can also be viewed in terms of central-peripheral organization. Trevarthen (1968) saw the ambient system as providing "the contextual information for the ordering of focal perception within the central part of the field." He viewed it as determining space at large around the body while focal vision examined detail in small areas of space. Orientation of head, posture and movement all occur as part of the ambient system as it serves to provide the surrounding context to the central detail (Bruner 1983). In an interesting analogy, John Streff, O.D., (1969b) compares ambient-focal action to that of a zoom-lens system while explaining the relative expansion and restriction of volume of the field of attention by a change in the x and y coordinates while maintaining a constant z axis set.

In earlier chapters, we discussed the effect on the stress response of constricting the periphery, reducing one's reception of external information. With this in mind, we can also hypothesize that the B2 case becomes a B2 because of the tendency to inhibit taking in full external information. This can be done by functionally constricting the periphery and by reducing awareness from the lower order ambient system. This constriction of periphery, the contraction of behavioral space, occurs primarily as a result of the feeling of vulnerability, similar to the way a one-celled animal contracts when threatened, and the feeling of vulnerability seems to be an increasingly significant product of urbanized civilized life.

It is interesting how one can account for the origin of the B1 and B2 cases types in so many different ways, all of them having a logical base. Ann Sutton Nichols, O.D., (1951) relates the B1 and B2 types to the positive and negative relative accommodation, plus and minus to blur-out at near (tests #20 and #21) rather than to the base-out and base-in prism tests (#16B and 17B). Charles Margach, O.D. (1984), in turn, relates them to the dark "locus" of identification and centering (dark focus and dark vergence), seeing the B2 type as arising when these dark loci are relatively far away, no closer than one meter, and the B1 type coming about in those having relatively close dark loci.

Effort and the Skeffington Model

Most models of vision and visual stress are effort models. They conceive of visual stress and discomfort as being directly related to some expenditure of effort in some aspect of visual function. Nearpoint stress is seen to come about because of the accommodative effort needed to focus at a close reading range, the vergence effort needed to converge for this near distance, and the additional accommodative and vergence effort needed to overcome hyperopia and exophoria. It is also seen to come about because of the need to use effort to reduce the lag or lead of accommodation brought about because of the biases inherent in the intermediate placement of the dark focus and dark vergence. Even in the physiological stress model, effort is usually thought of as being needed to restore accommodation to the plane of regard after the initial sympathetic activation as well as to block any induced overconvergence.

The Skeffington approach to visual stress, however, is primarily a non-effort model. If effort is involved, it is minimal and is the result of stress, not its cause. In the Skeffington model, a

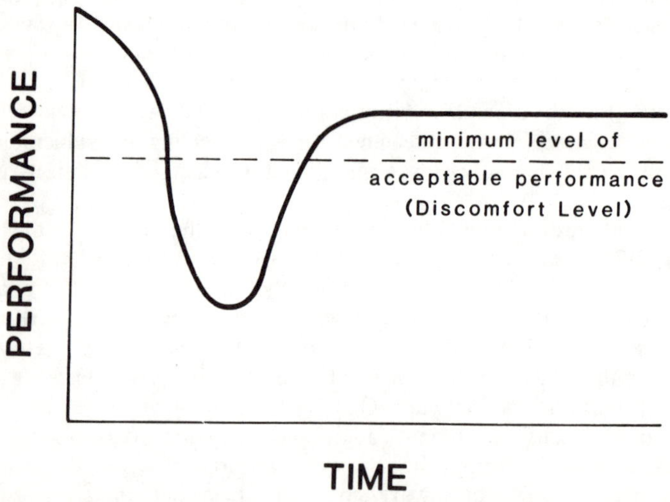

Fig. 31. Functioning above the discomfort level

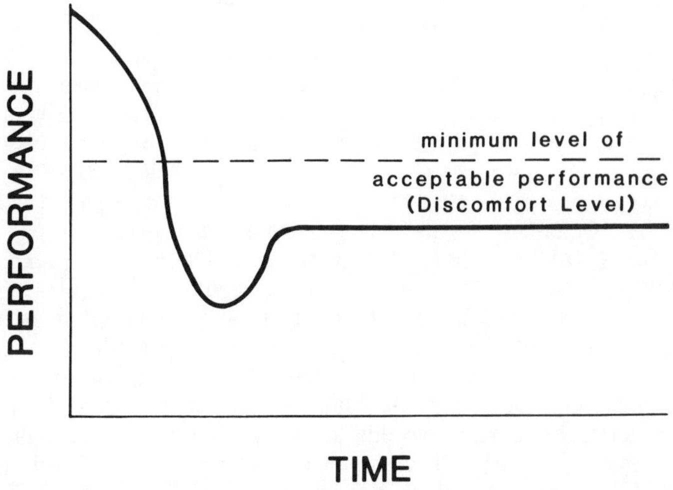

Fig. 32. Functioning below the discomfort level

slight lag of accommodation and convergence is normal and are homeostatic mechanisms to protect the system. They do not require effort. They are designed to minimize potential tensions within the system. Smooth muscle, visceral, autonomic systems are considered to be capable of maintaining function for long periods of time. Unless extreme circumstances persist, they are not considered to tire or fatigue too readily. Walter B. Lancaster, M.D. (1923), demonstrated that each of the extraocular muscles around the eye was at least one hundred times stronger than it needed to be to move the eye. Therefore, skeletal ocular muscles should not be fatigable under ordinary circumstances either.

The Skeffington approach deals with spatial performance rather than mechanical eye function, per se. Whatever happens to accommodation, convergence, identification or centering is viewed as occurring in space, on the z axis. Skeffington did not deal with the initial sympathetic stress response. If he did, he would have described it as accommodation or identification posturing further out in space. This initial response is countered, in at least the B2 case type, by parasympathetic stimulation. This is an automatic response. It involves no effort. It is part of the equilibrating action of the autonomic nervous system. The only effort that we might hypothesize is involved comes from the active cortical inhibition necessary to counteract the induced overconvergence. Pribram (Pribram & McGuinness 1975,1978), for one, noted that the major source of effort is inhibition. Skeffington (1937), however, did not even consider cortical inhibition as being effort producing. He felt that to inhibit, optometrically speaking, meant only "to readjust the mechanism as though the object were moved out." He saw this active inhibition as resulting in a perceptual constriction, a loss of flexible movement which caused a loss of efficiency and meaning. Fatigue was seen as being indirectly related to the loss of efficiency and meaning. Fatigue and discomfort are experienced only when outcomes are perceived as unsatisfactory to the individual, if goals exceed performance and the failure matters to the individual. Skeffington (1950c) quoted S. Howard Bartley, Ph.D., as stating that ocular discomfort occurs when achievement drops to a point unsatisfactory to the organism. Operationally, according to this concept, each individual has a minimum level of performance that is acceptable to him. As long as performance stays above this level he will tend to be symptom-free (see Figure 31). If performance falls below this level, discomfort, feelings of fatigue and anxiety as well as a loss of the drive or desire to sustain on the task results (see Figure 32). Discomfort, therefore, is not physiological. It is behavioral, "a complaint to headquarters" (Skeffington 1969).

The Skeffington model deals with function in space as it affects performance quality. The containment properties of prolonged nearvision tasks including intellectual demand, problem solving and the search for meaning are seen to result in a constriction and limitation of movement, learning, recall and performance: a loss of ranges of freedom. Containment, to Skeffington (1965b), produces this reaction, not the lag of accommodation and not effort. My own view is that the containment properties of a task are comparatively minor stressors and, of themselves, have a limited stress effect on the organism. The major stress producer appears to be one's attitude and mood: one's approach to the near-vision task that is superimposed onto the containment factor. The entire visual stress sequence, the way it develops and the choices that are made are determined by the pre-set of the individual, his mental state prior to and during his interaction with the task. It is this that upsets central- peripheral, figure-ground relationships and becomes the sign of a visual problem.

Regardless of what triggers the visual stress response, a key element of the Skeffington model has to do with the use of lenses. To Skeffington (1957c), lenses are not used to reduce effort or to correct for tensions due to the lag of accommodation, the position of the dark focus or an induced overconvergence. Plus lenses are seen to serve only one purpose, to minimize dissonance by satisfying the drive for convergence to posture closer than accommodation. The lenses simply allow accommodation (identification) to posture further in space than convergence (centering). By minimizing this dissonance in spatial positioning, it allows constricted movement systems to become more flexible. With increased movement and flexibility, overall performance, learning and recall improve. In this way, the plus lens satisfies the "let me get away from it" urge.

Chapter 22

The Time Factor

One general rule seems to hold true regardless of the particular adaptive path the visual stress response takes. The rule is that "a defective organism achieves ordered behavior only by a shrinkage of the environment in proportion to the defect." Although this was postulated by Kurt Goldstein (1939) to account for the functional behavior of those with severe physical or neurological injuries, it is valid for all kinds of adaptive behavior. Constriction of operational ranges goes hand in hand with adaptation.

All adaptive reactions occur in time. Some may take more time, some less. The very act of conditioning takes time and then becomes a way of bypassing time as one begins to react in the present based on prior ways of responding. When first exposed to a stressor (an event), most individuals adjust somewhat flexibly. With repeated exposure to the same or similar stressors, physiological changes occur that tend to become more structured. This is the basic process by which habits are set. During the initial stages of exposure, removal of the stimulus could reverse its effect and restore original function. When the stimulus is prolonged, when it occurs more frequently, when one's approach to it is more intense, when one does a task while feeling vulnerable and inadequate, the adaptive response becomes less reversible and more permanent. It becomes a learned reaction. As the adaptive response becomes conditioned, performance is not only affected locally, its effect is transferred to other areas of function that were previously not involved. As Goldstein (1939) put it, "the whole organism always participates in any reaction....with any change in one locality in the organism, simultaneous changes occur in other localities."

A simple case of a child with a broken arm may help illustrate this process. If a child has his arm in a cast for a period of one to three months, he could be expected to have some difficulty in moving the arm when the cast is removed. With renewed use and perhaps a few conditioning exercises, its function should be restored to normal. If, due to complications, the arm was kept in a cast for one or possibly two years instead of a few months, some residual degree of permanent limitation of function due to disuse might be expected. Even with an extensive program of rehabilitative exercises, the arm might never be capable of attaining its original quality and ease of operation. There may also be compensatory changes in the function of the other arm and in overall body posture as the organism attempts to re-establish a new functional balance on tasks involving the integrated use of both arms.

A similar condition exists in relation to a persistent visual stress response. The first area in which distorted performance is noted is that position in space most highly involved in the stress response--the nearpoint. In time, these adaptive effects are transferred to distance and become the basis for changes in refractive status. What happens at both distances is determined by whether the potentially stressful situation is avoided completely, partially resisted, passively accepted or engaged in strenuously and unflinchingly. It is also affected by the adaptive style that is taken. The two less common visual adaptive styles are generally quite stable: the one in which convergence collapses (see Figure 29) and the one where no attempt is made to compensate for the sympathetic stress reaction while convergence recedes outward to where accommodation is postured (see Figure 30). These conditions may improve but they do not usually tend to deteriorate much over time. This apparent stability seems to be noted in any form of adaptation where major concessions are made. Concessions of any kind are a way of escaping from a problem. They are avoidance (flight) reactions, preserving some function with a reduced level of internal struggle and readjustment.

The more common adaptive styles, the B1 and the B2 case types, usually show deteriorative or compensatory changes at near and far over time because there is a constant drive for internal readjustment to keep the functional components in some state of operative balance. When concessions do occur in these cases, they do not usually arise from internal physiological acts of flight but from an avoidance (intentional or subconscious) of the potentially stressful situation. There are many forms that compensation takes in these more common adaptive styles.

The first quantifiable response to a nearpoint visual stress situation occurs at near where the stress originates and results in the lowering of an exophoria, the normal homeostatic buffer, toward or into an esophoria, a tendency to overconverge. Prolonged or repetitive intense stress over time precipitates the following possible outcomes at this near area in space:

1. The original exophoria can be rebuilt usually to a higher degree to act as a better buffer to withstand the secondary effects of parasympathetic stimulation on convergence and to lessen the need for cortical inhibition of the induced overconvergence. This is an ac-

tive and positive attempt on the part of the organism to control the effect of the primary adaptive response to stress.

2. The induced esophoric (overconvergence) tendency can be maintained or increased to the point where a strabismus (esotropia) may develop. This response occurs when the organism is unable to rebuild a protective buffer to withstand the adaptive response to stress.
3. A higher degree of hyperopia can develop to act as an increased buffer to better absorb increased parasympathetic stimulation and basic shifts in neural dominance between the skeletal and visceral systems.
4. Myopia can develop at nearpoint (myopic projection). This occurs when one becomes conditioned to respond parasympathetically or when one exhibits a high convergence- accommodation/convergence (CA/C) response.
5. The changes that interfere with clear, single vision can be resisted with the result that performance levels decrease dramatically.
6. Function can be maintained by completely or partially avoiding nearpoint visual tasks.

A different set of possibilities occurs at distance. As was noted in chapter 19, according to the Skeffington model, at any point in space where one sees clearly and singly, one should theoretically be using one unit pattern of accommodation (1A) and one unit pattern of convergence (1C). At near, visually near- centered stress sets up a visual dominance which results in the activation of two unit patterns of accommodation (2A) to one unit pattern of convergence (1C). This induces the production of two unit patterns of convergence (2C) which must be blocked in order to maintain the two units of accommodation (2A) to one unit of convergence (1C) demanded by the visceral dominance triggered by the nearpoint stress situation.

At far, the organism is not involved in a primary visually stressful situation and, therefore, accommodation and convergence operate with one unit pattern respectively (1A-1C). When the 2A- 1C relationship at near is transferred to far, the following options become available:

1. The organism can reproduce the same conditions as exist at near (2A-1C) with the two unit patterns of accommodation at distance now producing myopia.
2. The same (2A-1C) relationship can be maintained but in a different proportion utilizing one unit of accommodation (1A) and reducing convergence to one-half

unit (1/2C). This results in a high exophoria or exotropia (divergence excess) at distance.

3. Both tendencies can be blocked maintaining single and clear vision at distance at the expense of achievement efficiency.
4. The original (1A-1C) condition at distance can be retained by an avoidance of stress-producing near-vision tasks.

Once the status at distance changes, compensations such as blurred vision (myopia) or an increased underconvergence (exophoria) tend to act as secondary stressors. Difficulty maintaining clear and/or single vision at distance then causes reverberations to be set up in which the former nonstressful area in space (far) starts transferring its effects back to the original stress-producing area in space (near). These near-far reverberations continue, depending on the demand for achievement at each area until some form of acceptable state of equilibrium is reached.

Factors Involved in Conditioning

The way an individual responds to stress tends to be conditioned over time so that it becomes quite predictable. An important element in this conditioning is how deeply embedded the response becomes. Embeddedness is an indication of the degree structure has changed to conform to the stress demand. It is always indicative of the progression being made toward restoring an operational state of equilibrium (Skeffington 1951c). A lightly embedded response is one where there are fewer changes in the organization of one's visual function. A deeply embedded response is one where there are greater and more persistent changes in visual function. Skeffington was able to isolate a syndrome, a cluster of findings, by which the state of visual embeddedness could be roughly estimated (see Table 9) (Skeffington 1951a,1951b,1970; Hendrickson 1980).

A second factor affecting conditioning is task demand or how one makes a demand out of a given task. The differences in response to varying levels of task demand have been a factor that has been noted for some time. This is seen in book retinoscopy. It is also seen in how often accommodative convergence/accommodation (AC/A) ratio values differ depending on how a test is taken. The difference in AC/A findings appears to be related to changes in the lag of accommodation brought on by the demand requirements of the task. The effect of low- and high-task demands on the lag of accommodation has resulted in the recognition of two types of AC/A measurements: stimulus AC/A and response AC/A. Yet individuals differ in how they respond

Table 9.

Syndromes of Embeddedness

The Non-Embedded Case

1. #4 (distance retinoscopy) lower than #7 (subjective--basic formula)
2. #9 (base-out to first blur at distance) excessively low, (in myopia #9 may be high)
3. #10 (base-out) recovery at distance excessively low
4. #11 (base-in) break and recovery at distance excessively low
5. #14B (fused crossed cylinder) net at near lower than #14A (unfused crossed cylinder) net
6. #16B (base-out) recovery at near excessively low in a B1 Case
   #17B (base-in) recovery at near excessively low in a B2 Case
7. #19 (amplitude of accommodation) and #20 (minus to blur-out) at near are both excessively low

The Embedded Case

1. #4 equal to or greater than #7
2. #9 up from its excessive low, stabilizing close to the expected value of 7 to 9 prism diopters
3. #10 break at distance low (below 13 prism diopters); the recovery finding is above the expected ratio of one-half the break measurement
4. #11 break and recovery at distance up from the excessive low; the recovery finding is in excess of the expected ratio of one-half the break measurement
5. #14B net equal to or greater than the #14A net
6. #16B and/or #17B break at near lower (below 13 prism diopters); the recovery finding is up above the expected ratio of two-thirds the break measurement
7. #19 and #20 are not both low; one is high while the other remains low (#19 is usually the higher finding in children)

to various demands. Skeffington (1951b) was aware of the importance of task demand over forty years ago in the taking of nearpoint crossed cylinder measurements. Crossed cylinder tests determine the flexibility or latitude of freedom in the identification process at nearpoint. The fused crossed cylinder test (#14B) is greatly different from the unfused crossed cylinder test (#14A). Even the respective phorias (#15A and #15B) taken through these findings are measured differently. Far more critical seeing is demanded of the individual in both the fused crossed cylinder test and in the measurement of its resulting phoria. These different tests (#14A, #14B, #15A, #15B) taken under different conditions often reveal different lags of accommodation.

Skeffington (1951b) realized however that the demand of a task consists of more than problem solving, immobilization and containment. It is also obviously more than something that is externally imposed. He sensed that the differences found in these various test findings reveal "the demands the organism is putting on itself to achieve understanding." Even at this early period, Skeffington recognized the importance of the psychological factor in determining how an individual responds.

A third factor affecting conditioning is the rigidity and the intensity with which an individual approaches not only a visual task but life itself. This rigidity is noted in everything one does. Skeffington (1970) attempted to quantify this factor diagnostically in terms of the "looseness" or "tightness" of one's spatial organization. The "loose" case tends to demonstrate unstable, unsure, variable and uncritical findings. More suppressions are noted and induced phoria changes are usually small (low AC/A ratio). The spatially "tight" case reveals more stable, unvarying and critical measurements. Little or no suppressions are noted and there are usually large changes in induced phorias (high AC/A ratio). The "tight" case appears to reflect a high level of uptightness, inner anxiety and a drive for accuracy that seems to stem from a need to prove oneself.

Embeddedness, demand and loose-tight spatial organization, therefore, are factors that affect and reflect how and to what degree conditioning takes place in the visual process and the changes that occur in the learned relationships between visceral and skeletal function.

The Stress Phoria

One of the things that I noticed soon after going into clinical practice was that the embeddedness factor was far from being a consistent indicator of adaptation to stress over time. Some very young children just starting out in school were seen to exhibit an advanced degree of embeddedness while many young adults who were avid readers and good achievers demonstrated a relatively low degree of embeddedness.

I also noticed that repeating certain tests at different times during an examination often resulted in slightly different findings. The general consensus was that this represented normal and accepted examiner/patient/instrument error ranges. This explanation was not totally satisfying because of other observations. Many patients who may have revealed shifts in certain measurements that were repeated during a given examination usually showed a consistency in findings from one examination to another. For example, an initial cover test in these instances would always match the initially taken cover test on a succeeding examination while a repeated cover test taken at the end of an examination might not match the finding taken earlier in the same examination. This phenomenon was also noted with phoria measurements and repeated crossed cylinder findings. Another interesting observation was that these changes in measurement, found on repeating certain tests after a short interval, no matter how slight they were, always appeared to change in the same direction in the same individual. One person would show a shift toward less exophoria or more esophoria on a repeated phoria test while another might reveal a consistent shift toward a greater exophoria or a lesser esophoria.

In light of my growing knowledge of the stress response, I soon began to view these observations as a dynamic probe into the real nature of what the embeddedness syndrome was only hinting at--the overall learned conditioned response to stress or how an individual visually reacted to a stressor based on his learned adaptive preferences. I (Forrest 1960c) settled on using a before-and-after Risley prism phoria test at near as the major element of a pilot study to investigate what I had begun to call the Stress Phoria. The procedure was done as follows:

At the end of a routine visual analysis, I would perform a standard Risley prism phoria test at a near distance (16 inches for an older child or adult and 13 inches for a young child), using whatever lens prescription was habitually used for close work. The patient would then be asked to read aloud slowly the paragraphs found in a standard nearpoint reading card, going back to the beginning when the last word in the last paragraph was reached. The aim was to give the patient a full five minutes of a concentrated, somewhat demanding, near visual task to do. If the material appeared to be too easy, the instructions were changed to read the card backward, starting from the last word or to read every other word, first forward and then backward, or to spell each letter or every other letter in each word, forward and backward. When the five minute period of concentrated reading was over, the near phoria test was repeated. Records were also kept during this pilot study as to whether the patient was a good or poor achiever, liked to read or avoided it, exhibited visual discomfort associated with close work, experienced a transient distance blur after doing close work or had become measurably myopic. If some degree of myopia was noted, it was thought to indicate that adaptation to near stress had existed for some time and its effects transferred to far.

The results of this study and subsequent investigations were very revealing. Nonreaders and poor achievers who avoided close work demonstrated the largest change (2 to 8 prism diopters) toward less exophoria or more esophoria. A high exophore who had a large shift toward less exophoria on the Stress Phoria indicated an overconvergence tendency that would be responsive to the effects of plus lenses at near. This was contrary to the classical notion that plus lenses increase an exophoria and are therefore often contraindicated when a high exophoria exists. This large change toward less exophoria or more esophoria revealed that these individuals were responding to each reading act as if it were a new stressor.

Those who did not avoid reading and close work but who tended to skim and take frequent breaks due to feelings of discomfort showed smaller shifts (0.5 to 2 prism diopters) toward less exophoria or more esophoria. Avid readers, especially those who had become myopic, revealed either little or no shift in the Stress Phoria measurement or a change toward more exophoria or less esophoria. What the Stress Phoria test signified was that the age of an individual or his basic phoria status was less important in terms of how he reacted to stress than a dynamic before-and-after measurement of his response to an actual visual task.

Similar revealing directional before-and-after changes were noted with the cover test, crossed cylinder tests and on selective aspects of the visual skills battery such as the Keystone visual skills cards DB9, DB4K, DB9B, DB5K and cheiroscopic tests, specifically figure tracing and the Van Orden star. It was also determined that a routine examination could be a stress in itself so that similar results could be achieved by doing certain tests early in the course of the examination and repeating them again at the end of the testing. Birnbaum (1985) has noted that overconvergence changes occurring in a near phoria test that is repeated rapidly approximately four times in a row can reveal a patient's sympathetic nervous system reaction to stress as could cheiroscopic tracing, if there is a drift toward an inward posture as the tracing is made. These procedures appear to be more of a probe of the initial Alarm Reaction of the stress response. The Stress Phoria, on the other hand, since it is taken both before and after a visually demanding task, appears to probe the individual's learned adaptive response.

In many respects, the Stress Phoria appears to represent an averaging of an individual's approach to life and how he interacts with his environment physically and emotionally. It reveals his time-oriented conditioned response to a visual stress. The initial "before" aspect of the test tells where he is now in time. The "after" aspect demonstrates the influence and power of past experience: where he has been, how he has previously responded to visual stress and how his past is influencing his present.

Chapter 23

Adaptive Outcomes: A Central-Peripheral Model

The nearpoint visual stress response as it functions over time results in a number of possible outcomes at both near and far distances. At near, one can adapt by becoming more exophoric, esophoric, myopic or hyperopic. Similar options are present when the effects of the near stress are transferred to far. The most common form of adaptation is an inhibition of major ocular changes at near and far at the expense of performance efficiency and comfort. Only certain people appear to become myopic while some others appear to have a tendency to develop major fusional instabilities. The basic Skeffington (1958c) model recognizes this but does not offer any concrete explanations as to who will take what option and for what reason. The only exception is a reference to myopia where the implication is offered that the person "has surrendered something, that he has self-limited his world." This suggests that the person is inwardly willing to sacrifice meaning-getting at a distance in return for something more important: achievement at near. It is true that many early myopes are not aware that they have a problem with distance seeing until it is pointed out to them. It is also true that many others are aware of, annoyed and unhappy by their distance seeing difficulty as soon as they become the slightest bit myopic. The term "self-limited" emphasizes the role of attitudes and mental states in the act of "choosing" ocular outcomes to visual stress. Although attitude may be a prime causative factor in the choice of these outcomes, there are still intermediate mediating factors that intervene between the hypothesized inner attitude and the end result. This chapter will pursue one of them: the role of preferred visual central-peripheral (figure-ground) processing styles.

Visual Central-Peripheral Organization

Throughout this book we have stressed that figure-ground organization is fundamental to life and to making effective choices. It has been noted that figure and ground are intimately interconnected. Figure contributes to ground and ground contributes to figure. Attention is simply the act of extracting a figure from a context in relation to the context (see chapter 6). Figure and ground organization is fundamental to how we perceive and conceptualize and to how we view ourselves in relation to others and to the surrounding environment. It is so basic to everything we know or do that many body mechanisms are uniquely designed to aid in organizing the world in terms of figure and ground. The visual process is one of these major bodily systems (Forrest 1976b,1976c).

The human eye has a very distinct anatomical construction to aid in figure-ground operation. More retinal receptors are concentrated in the central (macular) area than in the periphery of the retina. The neural transmission in the macular area is close to being in a 1:1 ratio of receptor cell (cone) to bipolar cell to ganglion cell (optive nerve fiber). In the peripheral retina, many receptors (rods) converge on a single bipolar cell and many bipolar cells converge on a single ganglion cell (Granit 1962).

Light entering the eye contains many potential units of information. Due to the neurological construction of the retina, the data from the light input can only be peaked for maximum detailed information processing in the central macular area. The surrounding retinal area processes the remaining information in a more or less diffuse manner to be used by the individual in less discrete ways. Each eye alone, therefore, provides a relationship between discrete and diffuse information, between a detailed aspect of the input which is selected by the individual to be "figure" and the nondetailed aspects of the input which become the "ground."

In a normal, two-eyed binocular situation a new element is added. The foveal areas of the two eyes are now aligned on the same point in space while the surrounding retinal areas of each eye are no longer in the same type of matching relationship with each other. The foveas have been traditionally called corresponding points, par excellence (Duke-Elder 1946). With the exception of the fovea and those corresponding retinal areas that make up the horoptor, all other aspects of actual space that are closer or further behind the object of fixation fall on noncorresponding points. This means that with fusion in play, the foveas are the best mechanisms for the union of similar or common elements while the remainder of the retina is best suited, under normal binocular conditions, for the analysis of uncommon elements triggered by differences in the positions of the objects in actual space that are not being fixated upon by the fovea at that particular moment. Out of this interaction emerges that unique attribute of binocularity--stereopsis or depth perception.

This process reinforces figure-ground relationships by adding similarity-difference discrimination to the basic discrete- diffuse discrimination inherent in the operation of each eye alone. By establishing such a variable detection mechanism, binocularity serves the purpose of better relating an individual to the different constituents of his space world and providing a means outside of actual touch by which different aspects of this space world can be related to each other.

It has long been observed that visual acuity tends to be better under binocular fixation. Electroencephalographic studies have verified this and demonstrated that with perfect binocular alignment, there is an actual doubling of the visually evoked cortical potential response (VEP) while the slightest deviation from perfect binocularity results in a purely monocular VEP response (Ludlam 1973). This demonstrates physiologically how the functional union of the discrete-diffuse and the common-uncommon systems adds an increased dimension not only to the peripheral aspects of central-peripheral organization but to central information processing as well. A binocular individual processes visual information quite differently than a monocular individual. Someone who is monocular can only utilize the discrete-diffuse mechanism in gaining information about relationships in space. In addition, those who have difficulty integrating these two different processing systems tend to modify their overall performance and ocular positioning in particular ways in order to adapt to the situation.

Nearpoint activities such as reading and writing that are done on flat surfaces work against the efficient integration of these two systems. Since these tasks are essentially two-dimensional acts,there is less noncorrespondence on the part of the myriad of stimuli impinging on the retina. This means that the common-uncommon system is being impressed with predominantly "common" information. The aspect of the total system that is geared to deal with differences now has to deal with similarities. How the system copes determines to a great degree the future visual status of the individual. Any deviance from a perfectly matched, well-balanced relationship in the interaction between the monocular discrete-diffuse system and the binocular common- uncommon system usually reflects a difference in the individual's style and quality of visual information processing.

Key Central-Peripheral Styles

Observations have indicated that the direction the visual process takes (e.g. esophoria or exophoria) in its conditioned adaptive response to stress is influenced by the preferred central- peripheral processing style. Both moderate-to-high esophores and exophores tend to demonstrate an inability to organize their central-peripheral relationships so that they operate efficiently as a simultaneous interacting system. In this respect they manifest the same basic difficulty. Their differences, however, are quite distinct.

The moderate-to-high exophore, whom we will call an exo processor, appears to want the best of both worlds. He has much difficulty operating his central-peripheral systems simultaneously as an integrated team yet insists on getting maximum information possible from both avenues. He alternates attention between center and periphery in varying ways as a given situation demands.

Though he can exhibit a central or peripheral style, depending on the task requirement, the exo processor appears to prefer to get as much varied, though not necessarily complete, information as he can per given unit of time from all areas of space. He prefers to operate as if he had many "de- peaked" maculas (or foveas) at one time. Since he wants maximum information, he may often tend to fixate off-axis in order to peripherally process as much of the field as he can.

When fusion cannot be maintained, the exo processor finds it more advantageous to posture his eyes outward in an attempt to maximize the ability to operate with many subdued simulated maculas with the least amount of overlap and conflict in processing peripheral information. Whenever fusion is lost, the common-uncommon system is switched off in favor of discrete- diffuse processing. In the case of the exo processor, the processing of the diffuse elements takes priority and the advantage of maintaining partial or total simultaneous vision is to utilize as much as possible of a two-field "diffuse" system as a compromise form of binocularity.

While the exo processor is "either-or" in his central-peripheral relationships, preferring a global perspective, the eso processor is highly central. The eso processor tends to be more efficient operating on a central basis and much less so when functioning peripherally. He appears to want less of a bite of the total information available while insisting on more depth and richness from the input that is accepted. He prefers a highly operative macula, monocularly or binocularly, in terms of information processing and will constrict periphery in order to achieve it.

When fusion cannot be maintained, the eso processor will find it advantageous to deviate his eyes inward, which aids in detuning peripheral information by having the peripheral visual fields overlap in a way that would result in conflict if peripheral suppressions did not take place. The two major styles of processing demonstrate that one could be a "central" or "either- or" processor depending on whether there is a predisposition for general information even though it is diffuse or more limited and specific information about one aspect at a time. Theoretically, at the opposite end of the spectrum, there should also exist a "peripheral" processor, one who would be very inefficient centrally. These individuals are in the small minority and for very good reasons. One of these is the act of accommodation.

Accommodation only occurs with central fixation. If the figure one picks out of the environment for meaning is not positioned within one degree of the center of the fovea, accommodation does not take place (Cornsweet 1970). The traditional graphs of retinal sensitivity show a high peaking at the area of the macula coming almost to a point at the center of the fovea, making the design of the graph look almost like a knight's lance. If this "lance" is pictured as an outstretched arm ready to direct attention on an object in space, accommodation can be considered as fingers reaching out to grasp and finger an object, while the process of identification serves to get the fullest meaning from it. In this respect, accommodation/identification is part of the "discrete" information processing system. This powerful meaning-getting reinforcement of the central alignment system makes it less likely that an individual who prefers to function peripherally will totally give up central processing.

A second reason is that it is far easier to depress functionally peripheral sensitivity, other than for survival purposes, than it is to eliminate central information completely. The powerful 1:1 neurology of the central system works against this happening. It is for these reasons that the exo processor tends to become "either-or" rather than solely peripheral in terms of information processing.

The central and peripheral systems serve different yet equally important functions. One is geared for the analysis of selected units of information. The other is less geared for detail. Since the peripheral system supplies more units of diffuse information per given "bite," it fosters a degree of predictability in order to get meaning from the array.

Those who prefer or perform better digesting one piece of information per unit of time and space tend to put less emphasis on peripheral information, using it basically for high priority survival purposes and as a general field for the selection of the next limited "bite."

Those who are good guessers, good predictors, good perceivers, who like to take "bets" on their accuracy in interpreting limited data rely heavily on peripheral information even though they do not necessarily neglect central processing. Language-arts tasks, for example, demand a high degree of interplay between central and peripheral information processing in order to achieve necessary detail discrimination with good ongoing predictive anticipatory processing.

This is the reason that those who function best in terms of peripheral processing do it in alternation with central processing. They manipulate both but have difficulty doing it simultaneously. When they are in situations that call for greater central-peripheral interaction, they often resort to an off- central processing that is typical of exo processors. Off-central processing is accomplished by fixating slightly off the object of attention while still getting meaning from the stimulus though the fovea is directed elsewhere. At other times, the exo processor might align directly on an object while looking through it. He might also depeak processing at all levels. When this is done, even diplopia is tolerated.

Some exo or eso processors demonstrate a high phoria or intermittent tropia at all distances, near or far, indicating that their preferred central-peripheral interaction mode is a consistent and pervading processing style under all circumstances. Some show it mainly at distance while others demonstrate it mainly at near. Still others may measure phorias within a normal range but have very limited vergence ranges. These differences are determined by factors such as the effect of accommodation which helps stabilize and reinforce central processing especially at near; whether close work is avoided or not; whether reading is done in short doses or is maintained; the priority for achievement; the priority for good spatial operation; the ability to utilize the disparity or noncorrespondence factors of binocularity; whether one is fixating a two-dimensional plane such as in reading and writing or manipulating three dimensional objects; the use of touch to support binocular function at near; and the demand for a specific function to be utilized such as stereopsis. Most exo processors, even if they are intermittent exotropes, will tend to be binocular in situations where there is a priority value for stereopsis. Regardless of the possible variations, the type of processing tends to remain fairly consistent for a given individual and is reflected in all performance areas. This means that the way an individual organizes his visual central- peripheral systems represents a total organismic action and not just a purely visual phenomenon. An "either-or" processor tends to exhibit a similar method of relating central and peripheral visual stimuli to each other as he does in relating central and peripheral auditory stimuli to each other. This is also reflected in the way a central visual stimulus is related to a peripheral auditory stimulus.

If these observations are projected onto personality styles, it could be postulated that an exo processor would tend to be generally aware of the presence of more than one point of view which could conflict with the need to develop a strong stand on any one aspect unless the demand was very great, while the eso processor would tend to be more generally aware of all that goes into a single aspect, generally minimizing the influence of secondary pieces of information.

The relationship between the phoria measurement and personality style is not new. Peter (1941) noted that esophoria was more prevalent in neurotic individuals and in those with a "highly strung nervous system." Attinson (1959) found that psychopathic subjects showed a higher incidence of esophoria than did a control group. Around the same time, optometrists Streff and Apell (1960) indicated that the exophore and exotrope are sensitive individuals who get a feel for a situation but are not able to decide how they arrive at a solution to a specific problem. They tend to be more intuitive. The esotropic individual is seen to strive for consistency, making rapid closure while seeking to get right answers. Skeffington (1972) viewed the exophore as relating a specific area in space to total space and the esophore as relating a specific area in space to himself. Birnbaum (1981c) found significant performance differences between esotropes and exotropes on the Children's Embedded Figures test. Male esotropes often demonstrated a more analytical, highly differentiated style than did male exotropes. This relationship was not seen among female strabismics, indicating a possible difference in cerebral hemispheric specialization.

There have been others who have noted the existence and the importance of central-peripheral processing styles though they were not specifically related to the phoria. Macdonald (1962- 1965), for example, described individuals with a central processing style as generally preferring to perceive in detail, in small bits, and constructing their visual space world sequentially through scanning. Those with a peripheral processing style were seen generally to perceive more globally, integrating large areas of space at one time. Neisser (1967) proposed that two distinct systems were involved in attention, one that analyzes many elements at once over a large area and one that focuses on fewer elements in a smaller area. Neisser's two-system hypothesis was supported in a study by Ambler and Finklea (1976). Eriksen and Rohrbaugh (1970) propose that attention involves a single system comprised of two aspects. They view attention as a variable power lens. Distributed attention is considered to be analogous to a low-power lens setting where many elements in the field are visible but where fine detail cannot be discriminated. In concentrated attention, the field narrows as with a high-power lens setting to provide an increase in resolvable detail. Fritz and Fehmi (1982) view attention in a similar manner "as a continuum with a narrow focus of attention at one end and diffuse attention at the other." What we (Forrest 1976b,1976c) have referred to as efficient simultaneous central-peripheral integration they (Fritz & Fehmi 1982, Fehmi & Fritz 1980) call "open focus," a style of attention "which admits both narrow and diffuse forms of attention simultaneously."

Central-Peripheral Style and the Z Axis To maintain concentration one has to inhibit the tendency to shift awareness. Potential stimuli are always around and the system is geared to be responsive to them so that choices can be made from out of this field. Maintaining focal attention for prolonged periods of time, forcing attention, can only be done at the expense of peripheral stimuli whose influence must be restrained, repressed or suppressed.

We have stressed only one aspect of central-peripheral (figure-ground) processing, that involving the organization of stimuli on the spatial plane encompassing the horizontal x axis and the vertical y axis. The exo and eso processor represented two styles of operation on this plane, each involving two processing modes, a monocular discrete-diffuse mode and a binocular common-uncommon mode. The only reference made to the near-far z axis was the emergence of stereopsis or three-dimensional vision out of the operation of the common-uncommon system. The z axis of space, however, plays a more important role, one that is highly involved with central-peripheral interaction.

If a close object is viewed, that object becomes figure to all the distant stimuli which form the ground in relation to it. If a distant object is viewed, the reverse is true. In other words, center and periphery relate as much to the action of a stimulus in regard to near and far space as they do to the interaction between a stimulus and side-to-side or up-and-down space. A global-focal style of processing information ("central" vs "either-or") is as much a part of near-far spatial figure-ground function as it is of a horizontal or vertical relationship to a central fixation point.

There is another set of factors involved in this process. These concern the making of judgments between objects in space and the making of judgments based on a personal relationship to what is out there. The first is an "it-it" function in which one computes relationships among various external spatial stimuli. It is how individuals localize and know where things are in space, what direction they are from the individual and how they relate to each other. This can be both an x-y and z- axis function, depending on where the objects are positioned in space. Harmon (1969) called this the process of localization. The second factor concerns what Harmon (1969) called orientation. This represents the ability to know where one is in space in relation to the things in space. It involves a basic assessment of the overall relationship between the individual and the objects, goal or task before him prior to acting on them. It is a "me-it" type of relationship between the individual and his goal that primarily involves the near-far (z axis) dimension using the individual as the base. This "me-it" function is more than just a referential system for making judgments in space. A major aspect of this process is attitudinal in nature, mainly whether the "me-it" is basically "me and it" or "me versus it." This happens to be a very important distinction that ultimately is a determining factor in the adaptive outcomes that lead to changes in refractive state.

The "me-it" function is related to how one views life. Since the body and bodily actions are symbolic of the mind and mental actions, how we judge ourselves in relation to our world will affect our central-peripheral preferences. The "me-it" relationship can be thought of as a behavioral continuum. At one end is the "me and it" dimension. "Me and it" represents a feeling of oneness with the world, being immersed with life and its experiences while feeling secure, protected and supported in all that one encounters. It is an operational mode that is associated with religious or mystic events, near-death experiences, love, intuition and creativity.

At the other end of the "me-it" continuum is the "me versus it" dimension. This represents a degree of remoteness or detachment from life and one's experiences due to a basic feeling of vulnerability. The inner feeling of vulnerability is at the root of almost all the adaptations we make. It can result in our becoming selectively aggressive. It can also lead to the suppression of certain areas of our interactions with life. Mental walls, blocks and resistances can be set up to preserve our illusions of ourselves. We can become reluctant to face certain aspects of life, holding back a little and avoiding making full commitments or becoming completely involved in situations. The feeling of vulnerability stems in part from the "existential dilemma in back of everything we do: this terrible anxiety we have about who we are and what we are doing on this planet, what it means to have our name and our face" (Becker 1982).

The "me versus it" dimension can be thought of as another bipolar continuum. At one pole is the "me against it" attitude. At the other extreme is the "it against me" belief. Those with a "me against it" belief system feel vulnerable and not in total control but react often by challenging and attacking the world. They project onto the world their own insecurities and feelings of low self-esteem. If they dislike themselves, they tend to project that dislike to others. If they find fault with themselves, they tend to be intolerant of similar faults in others. If they inwardly feel guilty about not living up to the standards they set up for themselves or that they perceive that others have set for them, they project that guilt onto others and try to make others feel that sense of guilt. They project anger and hostility onto other people and other situations. They fight what they perceive as authority when they feel threatened, oppressed or imposed upon. While the more healthy "me and it" people express ideas and emotions without being judgmental, the "me against it" individuals project ideas and emotions onto others. They figuratively push the world away from them. It is as if they confuse their inner world with the outer world, picturing the external world as having attitudes and actions that they refuse to see in themselves. "A person who is inclined to project, resembles someone sitting in a house lined with mirrors. Wherever he looks, he thinks he sees the world through the glass, whereas he actually sees only reflections of the unaccepted parts of his own personality" (Perls 1969).

Those with an "it against me" inner belief often tend to retroflect more often than they project. Retroflection means that attitudes and actions that should be directed outward are inappropriately turned inward toward the individual (Perls 1969). "It against me" people tend to become depressed easily. Depression is a form of anger directed inward rather than outward. They often tend to think more poorly of themselves, seeing themselves at a deep, possibly subconscious level, as being inadequate, guilty and deserving of blame. When they feel persecuted, put upon or oppressed, as they often do, they inwardly cringe rather than outwardly attack. If they attack at all, it is usually themselves. They are inwardized in all their activities often losing themselves in their work and psychologically losing a sense of time.

The terms projection and retroflection are not being used as substitutes for extraversion and introversion. There may be some overlap but there are decided differences as well. Extraversion and introversion refer to how an individual relates to others. The extrovert is more sociable, confident, outgoing, affirmative and expressive. The introvert is more introspective, quiet, thoughtful, hesitant and asocial, preferring the company of only a few family and friends (Jung 1976). Projection and retroflection refer to how the individual qualitatively views the world in relation to himself and whether these feelings are transferred inappropriately onto external things (people and events) or are magnified and inappropriately turned in onto himself.

The "me versus it" dimension, whether it is primarily "me against it" or "it against me," has other aspects. Many who feel some degree of alienation from the world may respond differently, depending on whether they are interacting with people or things. Some overreact to both. They react to people, social occasions, objects and natural events as adversaries. Living is simply a continuous battle. Others appear to react more to intrapersonal and psychosocial situations. Still others appear to feel more vulnerable in relation to things and physical challenges. They may be more at ease with people and less at ease during an electrical storm, when driving a car, flying in an airplane, crossing a bridge, reading a book, writing a letter, playing ball or walking on a two-by-four-inch walking rail. In any case, these are simply variations of the "me versus it" attitude. Knowing individual preferences helps in predicting what situation will induce a stronger stress response.

The reason for this discussion on "me-it" attitudes (see Table 10) is that clinical observations and discussions with patients seem to reveal that those with a strong and deeply ingrained "it against me" belief system are more prone to develop myopia while those with a deep-seated "me against it" belief system are more prone to develop a moderate degree of hyperopia. These are not the only factors involved, as will be brought out shortly, but they appear to be the major ones. The other factors simply determine if and when the specific adaptive outcome takes place and what its magnitude will be. The specific "me-it" attitude is the seedbed from which this behavioral transaction emerges and develops.

Table 10.

The Z Axis Dimension

I. "It-It" - Referential - (Localization)
II. "Me-It" - Referential and attitudinal - (Orientation)
- A. "Me <u>and</u> It" - Oneness with the world
- B. "Me <u>versus</u> It" - Detachment due to vulnerability
  1. "Me against It"
     a. Projection
     b. Hyperopia-prone
  2. "It against Me"
     a. Retroflection
     b. Myopia-prone

The Tension Dimension

Superimposed onto the various forms of central-peripheral organization and the preferred styles that different people manifest is the factor of tension. Tension represents the amount of psychological effort, intensity or resistance that is invested in a particular act. The tension dimension reflects how an action or movement experience takes place, how we attend, whether it occurs with total ease or maximum resistance--regardless of whether the task is physical, cognitive or emotional. Tension is related to the pressure we put on ourselves or that we perceive others put on us to perform or conform. Tension is cyclic in many respects.

Tension increases anxiety and is increased by anxiety. It induces a more narrowed focal, discrete, central processing style and is increased by the exclusion from awareness of peripheral stimuli. Tension also fosters selective rigidity in body function and in mental action. This rigidity or tightening up of body parts, what Whatmore and Kohli (1979) called dysponesis (see chapter 13), is triggered by tension and increases tension. Tension and rigidity are so interrelated that tension or resistance can be said to represent the rigidity in the way a situation is viewed or a task done. It is not uncommon for tension during a visual task to be reflected in a rigidity or tightening of the neck and shoulder muscles, increased staring (a reduced blink rate) and shallow breathing including a holding of the breath (Forrest 1984b, Dilts 1980, Mecacci & Pasquali 1980). Although these physiological signs of an increase in the intensity of visual attention can be found in myopes and hyperopes, they are more commonly seen in myopic individuals. The Fight, Flow, Flight And Fright Dimension

Another factor that is superimposed onto one's central-peripheral processing style is the overall path of adaptation that is preferred in general or is chosen in response to a particular situation. In chapter 3, these paths were referred to as catatoxic (fight), syntoxic (flow), flight and fright. In terms of a visual task or any major life event for that matter, these paths reflect diligent pursuit of the goal, effortlessly accepting and flowing with the experience, actively avoiding the situation (usually by the substitution of another activity such as daydreaming), and by simply not reacting at all (being immobilized, physically, mentally and/or emotionally).

Adaptive Outcomes--A Personal Model Four interacting factors were presented that come into play in any life experience, visual or otherwise. The first two relate directly to central-peripheral organization. The latter two act as modifiers. The first deals with how we get information from the world around us; how we select figure from ground and how we organize figure and ground in terms of discrete-diffuse and common-uncommon elements. The second is concerned with how we as individuals relate to and deal with the surrounding world on a personal "me-it" basis. The third pertains to the intensity or resistance we put into a particular endeavor which often is a reflection of how life itself is approached. The fourth, the path of adaptation, relates to how we cope with major life events, especially those that make us feel more vulnerable or those that we feel might adversely affect what others think about us or what we think of ourselves. Although these four factors stand as independent categories, there is some overlap. A diligent "fight" approach often goes along with a more central processing style, greater resistance to the task, and a "me versus it" belief system. In other instances, these relationships may differ. Any combination is possible. The reason for including these four factors in this chapter is that they form the basis of a personal model for viewing and understanding why and when certain outcomes occur in given individuals. It may not be a totally inclusive framework but it has been diagnostically useful, encompassing elements that previously have been enigmatic, not seeming to fit within the accepted conventional behavioral concept. The following are a few conclusions that are derived from this approach:

1. The syntoxic individual who "flows" with an event, who has a "me and it" attitude in relation to the world, who shows little tension and whatever tension is demonstrated is always appropriate and easily dissipated, and who has a flexible central-peripheral style, reveals a normal set of optometric measurements, full and flexible accommodative and convergence ranges and hardly any shift or a very slight shift on the Stress Phoria.

2. An avoider avoids because of a feeling of not being in control. He exhibits a "me versus it" attitude in regard to the world. Optometric findings will usually show a normal set of base measurements but will demonstrate a significant decrease in exophoria or an increase in esophoria on the Stress Phoria test. The avoider responds to each task he has to pursue as a new stress since he is being compelled to become focal when he would prefer to avoid the task in the first place. 3. An individual who exhibits excessive tension or rigidity when doing a visual task will have this tension reflected in lowered peripheral awareness and restricted and less flexible accommodative and vergence ranges regardless of the refractive status outcomes (hyperopia or myopia). A generally intense avoider may reveal an emmetropia or low-degree hyperopia but will demonstrate his inner tension in low vergence (break) findings and decreased positive and nega-

tive relative accommodation (plus and minus to blur-out at near) ranges. If this increased tension pervades the relationship between the visceral and skeletal systems of the body, it will be reflected in higher than expected accommodative-convergence/accommodation (AC/A) and convergence- accommodation (CA/C) ratios.

4. A functional moderate-to-high hyperopia seems to occur mainly in the presence of a predominant "me against it" attitude in which the individual projects his inner feelings onto the world. This appears to be the primary cause. If, superimposed onto this attitude, a person is a more diffuse rather than a central, focal or discrete processor, a hyperopia will be measurable in conjunction with an exophoria. If the person tends to be more focal or becomes focal when achievement is desired, he will reveal an esophoria in conjunction with the hyperopia. The eso- exo outcome reflects the central-peripheral processing style on the x-y axis plane. The hyperopia-myopia outcome reflects a "me- it" central-peripheral attitudinal style on the in-out z axis.

5. Functional myopia is seen to arise mainly in those people who have a fundamental "it against me" attitude or when this attitude is maximized under specific circumstances. Having this underlying core belief does not automatically infer that myopia will occur. It only sets the stage for it to arise. Without this primary attitude, other outcomes are more likely to emerge. The triggering agents for myopia are an increase in the intensity in doing a visual task, a more central processing approach with a de-tuning of peripheral information and a more diligent "attack" mode of operation. Although these factors have been recognized for some time as possible causes of myopia, what this model stresses is that even in the presence of these precipitating aspects, myopia will not usually occur as an adaptive outcome unless the individual has an underlying and fairly strong belief that he is vulnerable in respect to the world and tends to retroflect his behavior in relation to the environment with an "it against me" attitude. An "it against me" induced myopia is, in effect, a psychological and perceptual barrier set up by the individual to protect himself from the world. It is as if "out of sight" means "out of mind." If this particular individual has a strong "central" eso processing preference, he will likely become a myopic esophore. If he has a strong "either-or" exo processing preference, he will become a myopic exophore, demonstrating less exo or a slight eso when the situation requires him to be more focal.

Chapter 24

Adaptive Outcomes: The Role of Posture

Duke-Elder (1946) emphasized the relationship among posture, balance and vision in the human organism when he stated that "there is an intimate association between movements of the eyes and changes in posture," with a close functional reflex correlation existing "between the extraocular muscles and the labyrinths which record movements of the head in space, and the muscles of the head with respect to the trunk." He also postulated that "the power of projecting an object in space is based upon a dual mechanism made up visual and postural components: the object is projected visually with regard to the eyes and its projection upon the retina is orientated gravitationally by the postural mechanism which synthesizes impressions from the muscles of the eyeballs, the neck and the labyrinth." The implications of these statements formed the basis for many of the contributions of Darrell Boyd Harmon.

Harmon (1958) stressed that vision was more than a higher-order skill, that it was integrally related to lower-order gravitational mechanisms and that an individual's visual space world was essentially an extension of the organization of the body's gravitational system. He viewed the organism as being constructed of three basic spatial referential coordinate systems: the torso with its righting and counterbalancing reflexes; the head with its gravitationally-oriented vestibular mechanisms; and the visual system with the unified action of the two foveas (or a single fovea in the case of a monocular individual) acting as, what Henry Grossfeld (1951) called, a space center or a spatial referential benchmark. The neck was seen as being the transducer mechanism modulating the actions of the trunk in relation to those of the head and visual system.

Each of these individual coordinate systems was viewed by Harmon as establishing its own functional invariant axis. Operationally, these separate invariants integrate to form a composite organismic invariant axis around the y axis of the body. It is this composite invariant axis that becomes the base for the internal referential system that is projected out into space and through which one's visual world is structured.

To Harmon, the electrical discharge from the retina aids in establishing "basic support with gravity and also the alignment of the body." The action of the extraocular muscles aids in "determining direction of movement." And, the ability of the visual process to operate as a distance receptor beyond the reach of the body aids in "determining efficiency of movement affecting the total skeletal motor organization of the body."

Duke-Elder (1946) stated that "because of the close linkage of the visual mechanisms determining spatial direction of movement with other mechanisms determining posture, posture plays an important part in vision and in visual perception." Harmon went one step further. After noting the effect that head and eye movements have on posture and the effect that shifts in posture have on head and eye position, he concluded that vision and posture are not only highly interrelated but that all these functional interactions should be considered as being part of the total action of the visual process.

Harmon felt that the work and illumination demands of the culture force people into persistent and prolonged postural skews which result in muscular tensions that eventually warp the invariant axis. This results in spatial warps and changes in actual spatial operation ability. To Harmon, therefore, the way we organize, interpret and act in space is derived from the way we ourselves are organized in terms of our internal, gravitationally based, coordinate systems. Because of the extensive postural demands of the culture, it is warped posture that triggers most of the visual problems that are encountered.

This relationship was first noted by Lowman (1918) and Mills (1919) early in this century when they demonstrated that faulty posture could produce esophorias, exophorias, hyperphorias and anisometropias while efficient skeletal alignment in relation to gravity tended to reduce these effects.

Using these relationships as a theoretical base, Harmon proceeded to verify experimentally their existence and confirm their implications. His findings, both published and unpublished, include filmed studies and graphic records of respiration, blood pressure, heart rate, galvanic skin responses, visual reactions and muscular changes that occurred as part of his research. These indicated that electrical or mechanical stimulation to the neck or back muscles can cause visual changes that are observable and measurable with the retinoscope and the space eikonometer. Similar visual and spatial distortions were observed and duplicated in those exhibiting poor posture especially while engaged in nearpoint visual tasks.

Harmon found, through a study of body mechanics, that the most efficient posture for reading is such that the distance from the eyes to the book should equal the distance from the point of the elbow to the knuckle of the middle finger of the same hand. If this position is achieved and maintained, the body is in better balance with gravity with a minimum of distortion or pressure on the neck and back muscles. He demonstrated through electromyographic studies that adverse stimulations to the neck and back muscles occur if reading material is held beyond or in front of this mechanically balanced body position.

Harmon's studies also revealed that prolonged distorted posture can produce structural changes visually with a 0.6 to a 0.8 average correlation between the visual distortion and the body posture. When standing, individuals with astigmatism were found to tilt their heads laterally. Those who were myopic tended to incline their heads backward while hyperopes tended to incline their heads forward. Those with different structural focus postures in each eye (anisometropia) tended to rotate their heads out of parallelism with the x axis of the upper torso.

Those who tended to deviate their eyes outward (exophoria) tended to raise and rotate their scapulae obliquely and upward as if they were thrusting their arms vigorously out into space. Those who tended to deviate their eyes inward (esophoria) tended to lower their scapulae and rotate them obliquely and inward as if they were pulling their arms and shoulders backward and away from visual space. Even handwriting demonstrated a relationship to visual status with astigmats tending to elongate their words horizontally or vertically, depending on the type of astigmatism, myopes tending to write "small round letters with a vertical dimension of less than two millimeters" and hyperopes tending to write rounded letters "with a vertical dimension of over three millimeters."

These relationships were not always totally consistent. One enigma was the fact that not everyone who demonstrated a persistent postural warp always revealed a matching ocular effect. Harmon's first conclusion was that there is often a time lag during which period an organism acts "as if" a problem is present although it has not as yet affected structure nor is it measurable. When asked to account for those who demonstrated postural warps that persisted for decades while still not manifesting concomitant visual changes, Harmon (1966c) explained that the process could be affected in one of two ways. If a postural warp causes a visual warp, then the two will be geometrically related. A postural anomaly, however, may not always be responsible for an ocular anomaly. A persistent visual distortion can affect postural status. In this case, the geometry of the postural skewing will not be equal to the visual.

A second enigma Harmon was confronted with had to do more specifically with refractive status. Harmon felt that the alignment of the vertical segments of the body related to refractive status while the positioning of the lateral segments of the body appeared to be related to phoria measurements. He specifically associated a persistent shifting of the center of gravity forward toward the center of the near task with myopia. A shift of the center of gravity forward also tends to cause the head to angle backward (a chin up-and-out position) as part of the body's natural counterbalancing with gravity. Many individuals with a forward-placed center of gravity, however, exhibit a chin down-and-in posture instead. Harmon (1966a) interpreted this observation by stating that "somebody acting like a myope will move his center of gravity toward the center of his near task, but if he is trying to preserve some hyperopic latitude, he will also alter the position of his head and shoulders. The top will look like he is pulling away and the bottom will look like he is moving in." In other words, Harmon acknowledged that it was possible for head posture to override trunk posture in determining refractive status. What he did not pursue were the elements involved in this override, the implication behind the phrase "trying to preserve some hyperopic latitude."

In the previous chapter it was noted that Harmon (1969) contributed the concepts of orientation and localization. Orientation represents the ability to know where one is in space in relation to the things in space. It involves an assessment of the relationship between the individual and the people, objects, goals or tasks before him prior to acting on them. Localization involves locating the objects, goals and tasks, determining what direction they are from the individual and what relationship one object in space has with another. Harmon (1969) primarily conceived of orientation and localization as operational constructs. He felt that some individuals tend to be better at and rely more upon one of these functions than on the other and that this preference would, in many instances, differentiate the myope from the hyperope: the myope would tend to put localization ahead of orientation while the hyperope would do the opposite.

Harmon (1966b) only generally considered the physiological mechanisms involved in these functions. He viewed localization as being primarily related to the bilateral interaction of the body as it is involved in a task. Orientation mechanisms were seen as being far more complex, including the basic visual input (the retinal image); binocular function (the two eyes acting as visual range detectors); the stretch sensitive extraocular muscle receptors which measure the position of the eyeball in relation to the head; the neural messages from the cervical neck area (C2, C3 and C4) providing information as to the position of the visual system and the head relative to the body; the vestibular apparatus concerned with posture in relation to gravity; the body contact receptors; and muscle, tendon and joint proprioceptors (Cohen 1961). Of all these mechanisms, Harmon considered that the primary component relating most to the spatial orientation of visually fixated objects was the neural information coming from the neck area, C2, C3 and C4 (Harmon 1966b).

Robert Kraskin, O.D. (1981-83), whose philosophy of visual behavior is, to a great degree, based on the work of Darell Boyd Harmon, offers a slightly different view. To Kraskin, both orientation and localization relate to the task. "Orientation describes the resultant of the body's coming to balance with gravity and with the task." "Localization is a derivative of the body's coming to balance on the task and manipulating the task." In terms of body mechanisms, Kraskin proposes that orientation involves the muscles of the lower back while localization involves the upper back and neck musculature.

This model relates orientation and the status of the lower back muscles to refractive state. Changes in tonicity of the lower back muscles are seen to result in a shift of the center of gravity of the body either forward or backward. Myopia is considered to be "A reflection of body postural change resulting in the center of gravity moving forward of the norm." Localization is related to the "spatial computing status" of the body as affected by the musculature of the upper back and neck. Exophoria and esophoria, as noted in shifts in the position of the scapulae, are reflections of this process.

Consciousness and Posture: A Personal Model Continued

My own approach adds another dimension to the Harmon and Kraskin models. Where Harmon conceived of three organismic canonical referential coordinate systems of the body (the torso, head and visual system), my own philosophy considers that there is a fourth referential system, that of individualized consciousness which establishes an internal mental map with the ego as its invariant canonical center.

Some of the components (or "coordinates") of this fourth "mental" referential system are spatial (parallel) processing as it is counterpoised with temporal (serial) processing; sensory imagery as it is counterpoised with language; primary experiential processing as it is counterpoised with symbolic processing; central (focal) processing as it is counterpoised with peripheral (global) processing; logic as it is counterpoised with emotional feeling-tone; behavioral detachment mechanisms as they are counterpoised with behavioral involvement mechanisms, etc., all operating around an ego which serves the purpose of stabilizing the mental field by acting as a mental "space center" or mental "spatial benchmark" (to use Grossfeld's [1951] terms). The action of the ego, as we saw earlier (see chapter 15), is organized around attitudes, beliefs, biases and values through which the world is then interpreted.

It is through the interaction of these four referential systems, with consciousness at the helm creating an internal map that is projected out into space, that one develops the ability to judge one's self in space, relate to the things that are in space, relate things in space to each other, and to operate in space.

This internal map, biased by one's emotions, attitudes, values, beliefs, etc., also determines how one's physical mechanisms respond to situations including the flexibility or lack of it in returning to baseline normality either during or after an event.

Using this premise as a starting point, orientation and localization are viewed primarily as mental constructs and not necessarily as sensorimotor mechanisms residing in any particular body part.

Orientation is seen as the attuning or aligning of consciousness toward or with a stimulus, object or event, as a directional mental movement toward what the organism wishes to attend. It is an act of basic conscious recognition or interest that something is "out there."

Localization is seen as the process of establishing a relationship between one's self and what one wishes to attend; an act of mental triangulation between one's self as a bilateral organism and aspects of the environment. It entails a conscious "grasp" of what is recognized as being out there. In essence, both orientation and localization are seen as elements in the process of attention.

According to this view, orientation and localization at their most fundamental levels are considered as being mental constructs that are reflected in all aspects of behavior, visual as well as postural. In fact, it is how these constructs are organized around one's emotions and belief systems that influences who will act like a myope and become one and who will act like a myope and not become one but try "to preserve some hyperopic latitude" (Harmon 1966a) instead. In addition, by placing the primary mechanisms for orientation and localization in the fourth consciousness-based coordinate referential system and not in any particular group of sensory or muscle systems can we explain how an individual paralyzed from the neck down can still have adequate basic spatial orientation ability while exhibiting an inadequate or absent ability to move in space or manipulate objects.

This view also suggests that persistent refractive deviations and visual-spatial warps are, more often than not, reflections of pervading attitudes, emotions, beliefs, biases, value judgments and self-image. This is also true of rigid-type postural alterations. For example, it is a common observation that some individuals, regardless of age, demonstrate persistent residual visual and/or postural skews after completing a demanding nearpoint visual task. Others can apparently work for hours in adverse visual and/or postural situations and be capable of returning to a fully relaxed and flexible visual and postural stance when the task is completed. The key element appears to be due less to the potential "stressfulness" of the task than to the rigidity in one's approach to the task--one's psychological pre- set. The more rigid the approach the less the individual is able to maintain full and flexible operational degrees of freedom.

The rigidity in the approach to a task relates directly to one's orientation to that task. Keep in mind that orientation, as was discussed in the previous chapter, refers to the "me-it" relationship which is strongly attitudinal in nature: mainly, whether the "me-it" is basically "me and it" or "me versus it."

Luttgens and Wells (1982) recognize this aspect when they note that "atypical postures may be symptoms of personality problems or emotional disturbances." They consider that a hanging head and drooping shoulders may indicate shyness and a poor self-concept while an erect bantam cock posture may be an overcompensation for feelings of inferiority. Harmon (1966a) took note that people with a withdrawing makeup who say to themselves not "What do I mean to this out there?" but "What does this out here mean to me?" exhibit different physiological and postural patterns than do those who are more extroverted and outgoing. Observations of this type relating body tensions, muscular contractions, body posture and body usage to emotional and personality states are also behind William Reich's (1949) concept of body or character "armor," the "Alexander Technique" (Alexander 1969, Barker 1978), Ida Rolf's "Structural Integration" (Rolfing), and the "Feldenkrais Method" (1972). It is the psychological factor that appears to cast the major influence on all aspects of human function. In other words, the attributes of consciousness are the organizing matrix for most, if not all, of human and visual behavior.

Chapter 25

Adaptive Outcomes: Heredity, Use, Diet and Personality

Changes in refractive and binocular states are not universally accepted as being adaptive outcomes. Those who follow a genetic or biological approach consider them to be partially, if not totally, related to heredity. The classical view of the refractive state holds that an infant is born with a moderate degree of hyperopia which decreases with growth and age. If the infant demonstrates a large amount of hyperopia at birth, he will tend to remain somewhat hyperopic as he matures. If the child is born with a low degree of hyperopia, he will be more prone to become myopic as he grows older. These growth and age-related views account for the observation that most cases of myopia do not generally appear until the school years. This may be an adequate explanation for the onset of myopia in those between six and ten years of age but not in those who are over twelve years of age, by which time the normal growth of the eye should have been thoroughly completed. To account for the onset of myopia in this older age group, other followers of the genetic or biological approach theorize that refractive changes are related more to the genetically preprogrammed hormonal and metabolic effects of puberty than solely to age and growth factors (Angle & Wissman 1978).

The genetic theory is also weak in explaining the process of emmetropization. This is the name given to describe the fact that during development most eyes stabilize close to emmetropia to a greater degree than would be expected from pure chance. This process is accomplished through a complex interaction of the various refractive components of the eye so that as the developing eye grows and increases in length, the cornea and lens tend to flatten while the anterior chamber increases in depth so that the combined optics approach and maintain a state approximating emmetropia.

The genetic/biological approach has been popular for some time though it has always been in competition with a more functional, environmental, use/abuse approach. As with most theories, extremist positions rarely turn out to be totally accurate. The genetic/biological approach is obviously supported in those instances of children who are born with high refractive problems such as myopia and who also have parents and grandparents with a similar condition. It has also been related to premature birth, malnutrition, acquired systemic disease and ocular diseases such as glaucoma (Curtin 1970,1979). On the other hand, there appears to be a great deal of evidence to indicate that functional use of the eyes is of equal, if not greater, importance. Just as research in perception usually centers around visual perception, most investigations into refractive states usually center on the origin and progression of myopia. It just happens to be a more dramatic condition with patterns that appear easier to discern.

A Functional Approach

Ware (1913) noted that officers in the Queen's Guard demonstrated more myopia than did the more illiterate foot soldiers. This led him to conclude that literacy and reading were related to refractive state. Donders ([1864] 1972) observed that myopia was more prevalent in townspeople than in those who lived in the countryside. He concluded that it was due to the effect of prolonged tension when doing close work. In 1867, Herman Cohn (1886) found an increasing degree of myopia in school children as they progressed from the lower through the higher grades. He saw reading as being the primary cause but also attributed the type of school conditions that existed at that time such as poor printing in the school texts, desks that contributed to poor posture and inadequate illumination as being major contributing factors. He not only recommended good visual hygiene but encouraged the use of a head brace to encourage good posture when doing close work.

Young (1967b) noted in studies on monkeys and chimpanzees that two-thirds to three-quarters of the refractive changes that occurred were due to environmental conditions imposed upon the animals. Monkeys reared with visual-field-limiting hoods so that they could see only within a twenty-inch range developed more myopia than laboratory animals raised without hoods and decidedly more than those raised in the wild (Young 1961,1963a). He also found that, similar to human studies, female chimpanzees tended to develop higher degrees of myopia than males (Young, Leary & Farrer 1971).

Working with children, Young (1955) noted that the degree of myopia that developed appeared to be related to the amount of time the child spent in reading. The more time spent in reading, the more myopic the child tended to become. He also determined, as did Donders (1972), that the percentage of myopes was higher in urban as compared to rural populations (Young, Beattie, Newby & Swindal 1954) and, similar to Cohn (1886), that myopia increased with the grade level in school, tending to be about 5 to 15 percent in high school students and up to 20 to 40 percent in college students (Young 1965a,1977). Studying three generations of Eskimo families, Young (1969) found that close to 65 percent of the younger population were myopic while little or no visual problems were noted among the parents and grandparents. He attributed this to the younger generation being compelled to attend regular schools introduced into the culture in the last generation's era.

The relationship between visual function and refractive status has also been supported by the findings that higher ratings of bookishness are correlated with increasing myopia (Morgan 1960) that myopes are better readers (Young 1963b), that they demonstrate higher verbal scores on college entrance examinations (Young, Singer & Foster 1975) and have higher grade point averages than do either hyperopes or emmetropes (Young, Singer & Foster 1975; Beedle & Young 1976). On the basis of a regression analysis of over four thousand myopic eyes as well as a survey of the literature, Angle and Wissman (1978) conclude that "education (including reading and other close work) explains all the tendency for myopia to appear and progress through twelve-through seventeen-year-olds." A study of U.S. Naval Academy midshipmen revealed that 18 percent developed myopia by their senior year. All had 20/20 unaided distance visual acuity when admitted to the Academy (Hynes 1956). A more recent study also found a general and significant increase toward and into myopia in a population of midshipmen during the four years spent at the U.S. Naval Academy. Of six risk factors used for prediction, only the time spent reading as compared to the amount of time spent outdoors could be used to predict a shift toward myopia (Shotwell 1984). In a survey of five hundred second-year optometry students at Pacific University to which 89 percent responded, 75 percent of this select group, all of whom had completed nearly five years of college, had some degree of myopia (Septon 1984).

Myopia has also been related to working in a restricted spatial environment for long periods. This has been particularly evident in research on submariners. Schwartz and Sandburg (1954) reported on significant increases in myopia in those on extended submarine duty. Similar findings were reported by Greene (1970) while investigating submariners working on the Minuteman Launch Control Facility, by Luria et al. (1970), who found significant myopic and esophoric shifts among those on prolonged submarine patrol, and by Kinney et al. (1980), who concluded that "whatever factors are responsible for the increase in myopia in the general population are operating even more effectively among submariners."

Hypothesized Mechanisms

Most researchers and theoreticians who support the functional, use/abuse approach seem to agree that the most likely principal cause of changes in refractive status is accommodation triggered by nearwork. Some even feel that the developing eye grows toward emmetropia through the feedback that comes from increased accommodation to reduce the initial hyperopia (Wallman, Turkel & Trachtman 1978). Others have noted that through the use of near- vision lenses, bifocals or cycloplegia (atropine or scopolamine) the tendency toward the onset and progression of myopia is either decreased or stabilized in both animals and humans (Young 1967b,1965b; Bedrosian 1966; Kelly, Chatfield & Tustin 1975; Oakley & Young 1975). Gruber (1978) found an indirect relationship to accommodation when he noted that after discontinuing the use of atropine, myopia appeared to progress faster than in the control group and even faster than myopia normally progresses when atropine is not used.

There are two major theories in regard to the mechanisms involved in the formation of functional myopia. Both relate to the initial effect of accommodation. The first theory proposes that sustained accommodation causes a ciliary muscle spasm which leads to a hypertrophy of the ciliary muscle, making it more difficult to relax accommodation when looking at the distance (Sato 1957). The second theory suggests that a pressure gradient exists between the aqueous and vitreous chambers during accommodation (Coleman 1970) with prolonged accommodation causing an increase in vitreous chamber pressure which results in a stretching of the posterior shell of the eyeball so that there is a total increase in the anterior-posterior axial length of the eye (Young 1981, Bell 1980, Criswell & Goss 1983, Young 1975). Axial elongation has been found to be the most prevalent factor associated with increasing myopia (Baldwin 1981).

Not everyone, however, agrees with either scenario. Some relate the process of myopia formation to the position of the dark focus and the "lead" of accommodation (see chapter 20). Erickson (1978) implicates the depth of the anterior chamber as a major factor influencing the power of the eye. Anterior chamber depth is determined by the position of the lens with respect to the cornea and the retina. Young (1981) feels that the depth of the anterior chamber is of little importance in true axial-length myopia. Kaplan (1965) views the cornea as being a major factor in accommodation, proposing that the cornea undergoes a change in curvature due to pressure from the action of the extraocular muscles or from the pressure on the scleral spur triggered by the contraction of the ciliary body. Adel (1963,1966) hypothesizes that habitual strained accommodative efforts cause a traction of the choroid resulting in degenerative and atrophic changes which terminate in an axially elongated myopic eye. Donders (1972) believed that the increase in the axial length of the eye in functional myopia was due primarily to the pressure exerted by the extraocular muscles during convergence for close work and its corresponding effect on the fluid pressure within the eye. Greene (1980) supports this view by indicating that "the peak force exerted by the ciliary muscle is 0.6g whereas the peak force capability of the medial rectus muscle is 150g. On this basis alone, it simply must be true that convergence dominates accommodation." He also adds that "the oblique muscles can exert significant amounts of localized tensile stress on the posterior sclera (due to their insertions posterior to the equator and close to the optic nerve) and, with the increase in vitreous pressure associated with convergence, this concentrated stress may be sufficient to stretch the sclera out of shape permanently." Ku and Greene (1981) also note that "hard squinting or eye rubbing" can increase the intraocular pressure and cause irreversible stretching of the back of the eye.

This view is not too dissimilar in many respects to the concept advocated by Bates (1920). Bates proposed that accommodation is "entirely controlled (not by the lens but) by the action of the muscles outside of the globe," specifically the oblique muscles which lengthen the eyeball and the recti muscles which shorten it. He came to this conclusion by noting that many aphakics without lenses in their eyes could focus for various distances without either changing their glasses or using bifocals; that he himself could change his astigmatism at will; that Purkinje images reflected from the front surface of the lens were often variable during accommodation, sometimes becoming smaller and at other times larger; that some of his patients were able to accommodate to some degree after the prolonged use of atropine; and that "against" motion (myopia) could be revealed during retinoscopy when the patient was tense, anxious, fearful or was lying. Bates associated a change in refractive state with "straining" to see: myopia stemming from a strain to see at distance and hyperopia being created by the strain to see at near. Bates also proposed that the "strain to see" was primarily "a strain of the mind." He concluded that the remedy that leads to improved sight is not avoiding near or far activities but to get rid of the underlying mental strain which is reflected in an increased effort to see. This was to be done basically by relaxation. As Bates stated, "The fact is that when the mind is at rest nothing can tire the eyes and when the mind is under a strain nothing can rest them. Anything that rests the mind will benefit the eyes."

Functional Astigmatism

My own research into the origins of functional astigmatism (Forrest 1980b,1981a,1984c) has revealed that regardless of the primary cause, whether it is due to accommodative problems, convergence difficulty, presbyopia, poor bilateral organization, improper lighting, distorted posture, the mechanics of the task demand, the psychological desire for clearer vision over a wider range of space, or the mechanics inherent in using a bifocal, especially the narrower ones, one of the last mediating factors that must occur before astigmatism is formed is a change in the relationship between eye movement and head movement when scanning vertically as opposed to scanning horizontally. Horizontal (with- the-rule) astigmatism was seen to occur or be increased if the individual became more of an eye scanner (without head movement) when looking from side to side while yoking eye movements to head movements when looking up or down. Vertical (against-the-rule) astigmatism was observed to be triggered or increased when the individual became more of an eye scanner free of head movement when looking up or down while tending to yoke eye movements to head movements when looking from side to side. Obliquity of astigmatic axes and differences in astigmatism in one eye as compared to the other (anisometropia) were found to be brought on by persistent rotation, angling or tilting of the head superimposed onto the basic difference in meridional scanning style.

In the process of researching the causes of functional astigmatism, a number of interesting observations were made. Some hard-contact-lens patients who had difficulty stabilizing their lenses when they looked up began to move their heads more when gazing upward than when looking at the side. Within a few months they developed more horizontal or less vertical astigmatism. Soft-contact-lens patients whose lenses decentered when looking to the side due to narrow and tight lid angles were seen to become head movers when looking sideways while maintaining an eye-scan facility free of head movement when gazing vertically. Within a few months these individuals developed more vertical or less horizontal astigmatism. Many who retired, were assigned different jobs, changed their work or study habits (regardless of age or whether they wore glasses) were seen to have corresponding changes in astigmatism. Visual training programs stressing freer eye movements in the meridian where eye movement was yoked to head movement showed consistent reductions in astigmatism (Henderson 1934). These relationships between eye movement and astigmatism strongly implicate the action of the extraocular muscles as they affect the shape of the globe of the eye and/or the shape and positioning of the cornea.

In the course of investigating astigmatic status as it related to scanning habits, it became evident that it was not the meridian of free eye movement that caused astigmatism to form or increase but the meridian of restraint, the meridian in which eye movement was more consistently yoked to head movement. It was this meridian that became more myopic or less hyperopic. It was also noted that some individuals who were undergoing scanning therapy to reduce their astigmatism, decreased in astigmatism by becoming less myopic in the more myopic meridian while others became more myopic in the less myopic meridian. When these individuals were carefully observed while doing their assigned training procedures, the ones who became more myopic appeared to do the scanning activities more intensely, tightening their neck and shoulder muscles, staring more and even holding their breath. The ones who became less myopic appeared to do the procedures in a more relaxed manner. This resulted in changing the instructional set to stress a relaxed and effortless approach in doing the scanning procedures. Of more importance was that these observations reinforced the existence of a relationship between an increase in myopia and the intensity one brings to a visual situation.

The Role of Nutrition

Over the years there has been growing interest in the relationship of nutrition to myopia (Henderson 1934; Sorsby 1934; Knapp 1939; Wood 1927; Feldman 1950; Gardiner 1958; Vanacea 1971; Bardiger & Stock 1972; Lane 1981,1982; Koppel 1979-80,1983-84; Bowen 1984). Wood (1927) found that young myopes tended to have lower levels of free ionic calcium in their blood. This was also noted by Feldman (1950). Bowan (1984) hypothesizes that calcium spillage in myopes is "evidence of cortisol-induced hypoglycemia and the body's homeostatic attempts to correct the deficit by drawing on tissue stores of calcium. In the main, Bowan feels that nutritional imbalances are usually stress induced.

As a result of an extensive study, Lane (1982) points up the importance of three related deficit-inducing diets associated with the development of myopia: (1) "excessive consumption of refined carbohydrates in too high a ratio of consumption of complex carbohydrates, (2) excessive consumption of denatured flesh protein, and (3) inadequate intake of dietary fiber and folate."

Lane indicates that refining carbohydrates results in the depletion of chromium. Too high a diet of refined carbohydrates (sugar) leads to a chromium deficit in the body. Excessive consumption of flesh protein leads to a spilling of calcium into the urine and hair tissue, resulting in a deficiency of calcium in bone and other body tissues. A reduction of calcium from the sclera of the eyes could make the eye more susceptible to being deformed in response to intraocular pressure changes and to eyelid pressure. In addition, excessive consumption of cooked protein also reduces the uptake of food folate, pantothenic acid and pyrodoxine.

In terms of myopia, this study revealed that myopic patients, especially those with increasing myopia, had the lowest concentrations of chromium in hair tissues especially in relation to vanadium (Lane 1981). Myopes, ages seven through seventeen, had the highest hair concentration of calcium. Young myopes were statistically found to consume more than triple the ratio of refined carbohydrates to total carbohydrates as compared to hyperopes. Young myopes, not increasing in myopia, had a significantly lowered distribution of flesh protein intake compared to those becoming relatively more myopic. Young myopes were found to ingest daily a mean of 5.58 grams of dietary fiber while young hyperopes, on the average, consume 15.43 grams per day.

Lane concludes by stressing that myopia prevention requires the adequate ingestion and absorption of nutrients that aid in the maintenance of normal intraocular pressure levels and that build scleral collagen strength so the sclera can resist distension. Bell (1978) notes that collagen makes up about 70 percent of the dry weight of the sclera and emphasizes the importance of vitamin C in the metabolism of collagen. Rotman (1974), in fact, recommends using large doses of vitamin C in the treatment of myopia. Chvapil (1967), on the other hand, points out the enhanced effect that vitamins A and E have on collagen formation.

Young (1969) noted that the younger, school-attending generation of Eskimos was significantly more myopic than either of the two preceding generations. Although he saw nearwork as being the major causative factor, he cited the following passage from Elizabeth Cass' presentation to the Twentieth International Congress of Ophthalmology in Munich, August 1966, to indicate that there is more than one way to look at the subject:

Myopia is unknown among pure-blooded Eskimos. The majority have negligible refractive errors and a small number have low hypermetropia. Children whose parents have no refractive errors, after living on the white man's food for some years, develop myopia and carious teeth, i.e., they change from a high-protein to a carbohydrate diet. That diet does have some effect on the production and increase of myopia is confirmed by the work of Walkingshaw (1964) and Gardner (1964) in England (Young 1970).

The Role of Personality

In the previous two chapters adaptive outcomes were viewed in psychological terms, emphasizing how people differ in how they process spatial information and in the way they relate attitudinally to the world. This type of psychological approach is not new. Thorington (1900) observed that myopes were introverted, shy, had fewer friends and preferred indoor activities. Rice (1930a,1930b) reported that: A near-sighted child...is not dependent on others for entertainment and is liable to grow rather contemptuous of the abilities of others. He does not adapt himself to his surroundings and is not willing to make compromises. He is often severe in his righteousness and his rightness and may become a disagreeable personage." A hyperope, in turn, is "...nearly always a jolly good fellow....He has a ravenous appetite because of his activity, he scarcely knows fatigue...He is tanned, masculine, very aggressive, and is likely to be a devil with women.

Dunbar (1947) viewed myopia as a defense mechanism used by introverted children to protect themselves from the external world. Mull (1948) noted that myopes and hyperopes seemed to have distinct personalities with myopes being more overcontrolled and introverted and hyperopes being more impulsive and extroverted. Gesell (Gesell, Ilg, Bullis & Getman 1949) observed that myopic children appeared to exhibit marked tendencies toward introspection and introversion. Using the Guilford-Martin temperament test, Schapero and Hirsch (1952) found that myopes demonstrated significant tendencies toward overcontrol of the emotions, inertia, a strong need for social dominance and a disinclination for motor activity. Van Alphen (1952), in the same year, also noted that myopes had better control of their emotions, were concerned less about the outside world and thought more in abstractions. Stevens and Wolff (1965) found that myopes tended to be "sharpeners" while emmetropes and hyperopes tended to be "levelers." Using the Rorschach test, Rosanes (1966,1967) noted that myopes were more cautious, doubting and compliant with a high tolerance for anxiety. In a later paper, Rosanes (1970) observed that "the type of anxiety prevalent in the myope is one that he generally does not experience; it is covert, with a decrease in motor activity. It might be said that he is afraid movements might be seen and be dangerous and so he tends to hold back activity and thus frustrates himself." Young (1967a), using the Edwards Personality Preference Schedule, determined that myopes had higher needs than nonmyopes in the following areas: achievement (the need to do one's best and to be successful in whatever task is undertaken), introspection (analyzing one's motives and feelings, to observe others and to put oneself in another's place), succor (the need to have other people provide help when in trouble, to seek encouragement from others), abasement (to feel guilty when one does something wrong, to accept blame when things do not go right), heterosexual (going out with a member of the opposite sex, engaging in activities with the opposite sex), and aggression (attacking contrary points of view and telling others what one thinks of them). Nonmyopes were found to be higher in deference, order, exhibition, dominance and the need for change. Comparing these findings with a study of cadets at the Virginia Military Institute done eight years later, Young (Young, Singer & Foster 1975) found that only two main variables actually distinguished myopes from nonmyopes: heterosexuality and achievement.

Kelley (1971) pointed out a difference between the basic personality patterns of myopes and the various compensations they make in coping with the environment. He viewed the basic myopic posture as one of fear that is compensated for by stubbornness, determination and anger. The myope blocks the feeling and expression of fear while the hyperope blocks rage. Lanyon and Giddings (1974) reviewed the psychological approaches to myopia and concluded that "it has been consistently shown that myopes tend to be, in comparison to nonmyopes, somewhat more introverted, overcontrolled and tolerant of anxiety." Beedle and Young (1976) found that myopes described themselves with adjectives such as ambitious, conscientious, fair-minded, informal, kind, mannerly, pleasant and generous. Hyperopes were most likely to select adjectives such as anxious, changeable, defensive, excitable, impatient, nervous, stubborn and worrying. Streff (1977) in a study of school children, found that as children "develop anxiety related to tests or doing well in school, they are more likely to make a myopic adaptation to help them in that task." Zeiger (1976) found that moderate myopes view the world as repugnant, tend to reject it and withdraw from it. Low myopes show a stronger aversion, high myopes less. To Zeiger, "Low and moderate myopes show the problems and attitudes myopes have and the high myopes show solutions that myopes have." Singer (1978), in a doctoral study, noted that myopes are predisposed to work and play independent of the group, valuing wisdom and independence. He also was aware that there are some individuals who are able to achieve quite well in an academic environment without ever developing myopia. Gawron (1981,1983) sees myopes as exhibiting sympathetic rather than parasympathetic dominance. These views on the relation of personality to visual status led Young (1981) to conclude that "it is possible to identify quite accurately those potential myopes and nonmyopes by observing behavior and determining the values and goals of the children. Myopia is related to behavior."

Finally, a word about personality and strabismus. In a comprehensive review of the subject, Groffman (1978) offers two perspectives. First is the observation that many visual problems may be the cause of psychological or personality problems. Visual difficulties may trigger anxiety because of embarrassment, the fear of blindness, the fear of abnormal vision, the fear of loss of a body part, the view of vision or the eyes as symbols of other organs such as the genitals with the guilt feelings that may be associated with them, and the fact that children sense that their parents are disturbed by the condition (Berman 1976; Hart 1949; Heaton 1968; Lipton 1970; Lion, O'Neill & Prager 1943; Pollie, Haffner & Krasnoff 1964).

Secondly, Groffman cites studies and opinions that indicate that those developing visual disorders tend to be among those "most likely to display personality patterns similar to people with known conversion symptoms" (Mucha & Reinhardt 1970); that 21 percent of all strabismus is of psychological origin, especially those that appear after eight years of age (Pugh 1934)); that strabismus "should be considered as a developmental, maladaptive, psycho-physiological distortion, part hereditary and part environmental like other psychosomatic conditions that originate in infancy" (Beckwitt 1971); that at an ophthalmic clinic and mental health center in Lisbon "strabismus is considered as a psychomotor symptom" (Moutinho & Dos Santos 1968); that psychogenic factors such as anger and spite are involved in the etiology of strabismus, with unresolved anger being related to convergent strabismus which is interpreted as a "tightening of the ego defense" and an attitude of apathy and resignation being related to divergent strabismus, a condition interpreted as a "giving up of the defense in a reality situation which appears hopeless" (Rappaport 1959); and that "a child with a latent strabismus may develop a manifest strabismus after stressful episodes as a reactive defense mechanism, a method of acting out, or as a means of manipulating parents" (Allen 1965).

Chapter 26

Adaptive Outcomes: The Ann Sutton Nichols Approach

As we have seen, many people have been interested in the psychological aspects behind visual problems. One of the major contributors to this approach has been Ann Sutton Nichols,O.D., (1943a,1943b,1943c,1945a,1945b,1946a,1946b,1947,1948,1950,1951; Tramonti 1984). Unfortunately, most optometrists today are unaware of her work. Many of those who recognize the name associate it solely with the AN Series of Keystone visual training cards that she designed and which carry her initials. Ann Sutton Nichols, however, was an original and insightful optometric thinker whose ideas were firmly rooted in the psychological and behavioral correlates of visual behavior. This chapter, devoted to briefly summarizing a few key elements of her philosophy, is derived from available literature and transcripts.

In the conclusion to her 1946 Year Book, Nichols states: "This work is dedicated to and presented as a gift to the profession of optometry. The only price put upon it by the author is that it should be presented in pure form, unadulterated, undisguised and unpolluted by attempts at preemption."8 With the exception of some explanatory remarks and hopefully judicious editing, most of the contents of this chapter will be in her own words. In this manner, Nichols' legitimate request will be honored. If gaps or inconsistencies appear evident, it is most likely due to the fact that only extracts are being presented. In addition, the transcripts currently available are only a small portion of her total lecture and seminar output.

Basic Principles

One of the major principles in Nichols' work relates to the individual's visual space world. Nichols conceives of behavioral space as a cube and postulates that though space is three- dimensional we perceive it operationally as if it were in six dimensions. The first two dimensions are x-y frontal planes that are parallel to the width of the observer's body and perpendicular to the horizon. One of these planes is close to the observer and is called the "plane of identity." The other is further out in space and is simply called the "plane opposite the identity." This plane represents what is observed "out there" in space, whether objects or people.

Nichols also relates space to behavior by labeling the closer frontal plane and the further plane as the WHICH and the WHAT dimensions. The planes that form the sides of the cube represent the depth dimension of space. The one to the left of the observer is called the HOW dimension while the one to the right is labeled the WHY dimension. The horizontal planes that form the top and bottom of the cube are called the WHEN and WHERE planes respectively. (Unfortunately, in the existing transcripts these designations are not sufficiently defined.)

At another level, Nichols describes the emotions as representing the WHO dimension or the identity. WHAT we think and HOW we act is predicted upon WHO we are emotionally in relationship to WHAT we receive from the surrounds. In other words, the identity, the plane parallel to the self and perpendicular to the horizon, is the plane of the emotions. She hypothesizes that the direction of emotion is always in and the direction of thought is always out.

She feels the reason we become confused between ourselves and the surrounds is that we are so often mixed between the dimensions of WHO and WHAT which are at the level between emotion and sensation. She notes that some people tell us that they see space move in relationship to them instead of that they move in relationship to space. She conjectures that it is the imagined necessity for power over the surrounds that deludes us into shoving space around to suit ourselves. We either push space toward us or away from us rather than leaving it where it is. Actually, space is as it is regardless of how we interpret it. The distortion of space occurs because we tend to conceive it rather than perceive it.

A second principle of the Nichols philosophy relates to the innate versus the learned, the genetic constituents versus acquired behaviors. Nichols uses the terms temperament and case type. Temperament is innate. It consists of what we are born with constitutionally and temperamentally. The case type is what is acquired early in life based on behavioral interactions. It is how the innate predispositions are molded to cope with situations that are encountered in the course of living. The refractive state emerges from the innate factors of constitution and temperament as they are reflected in the acquired case type.

A third principle relates to the diagnostic components of the four major case types: the B1 myope, the B2 myope, the B1 hyperope and the B2 hyperope. Unlike the Skeffington model which determines the B1 case type on a low base-out vergence finding at near (#16B) and a B2 case type on a low base-in vergence finding at near (#17B), the Nichols model bases them mainly on the nearpoint minus to blur-out (#20) and plus to blur-out (#21) and secondarily on the blur component of the base-out and base-in at near vergence tests (#16A and #17A). She views the minus to blur- out finding (#20) as revealing how the person interprets from disparate points in space toward himself while the plus to blur- out finding (#21) reveals how he interprets from disparate points in space away from himself. When relating these measurements to the myope and hyperope, Nichols makes the behavioral distinctions noted in table 11 with the understanding that one can be a case type and yet not demonstrate a refractive "error."

Table 11.

Case Typing: The Nichols Approach

B1 - Myope

#20	very high
#16A	low, running into exo

The myopia is in the total organism. He is not seeing outside, in the world, but inside himself.

B2 - Myope

#21	high
#16A	high, tending toward eso

The individual looks down his nose, passing opinions as to his world, evaluating everything in it, according to his standards.

B1 Hyperope

#20	high
#16A	low, running into exo

The individual impresses the audience to aggrandize the self.

B2 - Hyperope

#20	low
#17A	low, with eso or low exo

The individual projects toward his horizon in service and with procrastination as regards his own goals.

In a further elaboration of the behavioral aspects of spatial planes, Nichols notes that some case types do not have the height dimension in visual consciousness, or if they do, it is very diminished. Some do not have the horizontal dimension, or if they do, it is not pronounced and therefore the field appears collapsed. The way different people compress the field, or segment for space and for objects in it in any of the three dimensions or the six cardinal directions, is done in accordance with their case type (not necessarily their refractive state). If the #20 finding is higher than the #21, it indicates that the horizontal dimension is diminished. The difference between these findings is the extent to which it is diminished.

Since man tends to view himself as the prime mover, force and power in his space world, he tends to identify the self in a position perpendicular to the horizon or the surrounds around a vertical cleavage. Both the B1 and the B2 case types have vertical cleavage traits. To Nichols the B1 prefers to view the self through his own motives and therefore identifies the surrounds with the self. The B2 prefers to view the surrounds through his own motives and therefore identifies the self with the surrounds. For this reason, she feels that the B1 selects to view the world and objects in it in the direction of a plane perpendicular to the horizon and parallel to himself, for he prefers the parallel perspective in that he wishes to identify the surrounds or horizon with the self. For the same reason, the B2 selects to view the world and objects in it in the direction of a plane also perpendicular to the horizon, but at right angles to the self, for he prefers the right angle perspective in that he wishes to identify the self with the surrounds or horizon (see Figure 33).

The B1 myope has a bidimensional spatial view of length and height without depth as he projects from the horizon toward the self. In order to fixate in the direction of a plane parallel to the self at all distances (parallel perspective), the B1 myope becomes more exophoric as myopia progresses.

The B2 myope has a bidimensional view of height and depth without length. He also channels his energies in the depth dimension, projecting in a direction from the surrounds toward the self (the realm of emotions). Since he prefers to fixate in the direction of a plane at right angles to the self at all distances (right angle perspective), the B2 type becomes less and less exophoric as myopia increases.

Nichols repeatedly stresses that not all myopic patients become esophoric during incipiency and exophoric as myopia progresses. She hypothesizes that the patient becomes more exophoric as he

Fig. 33. Refractive styles and spatial preferences

views a perpendicular plane parallel to the self and becomes more esophoric as he views a perpendicular plane at right angles to the self.

Both the B1 and B2 hyperope have a bidimensional spatial view of length and depth without height. They both select a horizontal cleavage and therefore a horizontal plane rather than a vertical spatial cleavage and a perpendicular plane.

The B1 hyperope is seen as maintaining tridimensional vision at all points within his farpoint although he may intermittently sacrifice binocular vision to maintain it. The B2 hyperope is seen as maintaining tridimensional vision at all points between his near- and farpoints and also being willing to sacrifice binocular vision intermittently to maintain it.

Each individual, therefore, as a case type and temperament relates himself in space in a different mode or manner and each selects a different plane or planes to which he projects his visual impressions. He positions the plane or planes in relationship to the self, according to his motive and personality. Lens and prism measurements simply reveal how he structures his space world.

The B1 Myope

As a temperament, the B1 myope is very passive and impressionable. He functions primarily at the emotional level, "feeling" his way through the world emotionally. He takes emotional impressions of everything he sees, hears and feels. He looks at people to see how they feel, not how they look because the way they feel affects the way he feels. He relies more on impressions than on interpretation. He is less interested in what you say than in how you say it and how you feel when you say it. He is loving and pleasing in order to preserve emotional peace but can easily become resentful when he submits to other people's ways for the sake of peace while really not liking to do so.

As a case type, the B1 myope neither knows nor cares what others think about. He does not get their ideas, but does know how they feel. He can perceive their feelings through his feelings, but he cannot think in their realm. His point of contact is through the world of emotions and not the world of ideas.

The B1 myope compresses his world in the depth dimension and ignores the horizontal plane, particularly beyond his farpoint. Consequently, when he walks through space, his horizontal plane unfolds before him as he walks, like a carpet, and until he takes a step, or feels the horizon under his feet, he does not recognize the horizon or have any realization of it. He does not perceive a great deal when he rides as a passenger or sits on a park bench as an observer. For the minute he sits down and is not in action, he begins to examine his inner processes.

The world as seen through his eyes is fuzzy, enlarged and compressed toward him...close enough to feel it but ever so remote for reasoning in it and utilizing it. Though the reasoning process is blocked, he is none the less very intuitive and intelligent, often accurately feeling the essence of things that cannot be perceived with the reasoning process.

In order to please, the B1 myope has a pleasing personality and a gracious manner. However, it is furthest from his intent to please anyone, even himself. As a child, he is socially dependent. He becomes a master at manipulating people and situations and when coerced, may distort truth and facts to suit his own intents and purposes. In appearance, his hair may be tousled and his clothes are often the casual or sportswear type, casually worn, fairly clean but wrinkled and disarranged on the figure. He does similar things to his surrounds. The degree of myopia reveals the level to which these attitudes are ingrained in his psyche.

The B2 Myope

As a temperament, the B2 myope is a very logical person. He belongs to the intellectual group, analyzing everything that goes on and making a sincere attempt to be analytical and accurate. B2 myopes think more than they see. In fact, they think they have more knowledge than most people. They are naturally a little timid. They never verify anything at the emotional level. They are their own best critics, having a highly stylized view of everything from an exalted position. They conceive of things rather than perceive them.

As a case type, they are proud and logical. They want an opportunity to accomplish things. They do not want sympathy but they do want a dignified, sympathetic manner. They glorify themselves and want an appreciative and understanding audience. Where the B1 myope prefaces his statements with "I feel...," the B2 myope begins by saying "I think...." The B2 myope's reasoning processes are highly developed. He is very keen and has a good intellect but his emotional processes are almost completely blocked. He loves you but will not express it as the B1 myope does by embracing you, but rather he must keep you in his central field at a respectful distance where he can evaluate you. Since he compresses his field in the vertical dimension and toward the horizon, he cannot see you in relationship to others but only to himself, for you and he are in the same line of vision and no peripheral field exists for him at that moment.

B2 myopes are baffled that they should remain obscure when they feel they are most obvious in their compassion and their offering. They are the perfectionists who demand discipline of themselves. They are not always equal to understanding the motives of those who have less compassion and who attempt to command more perfection of them than they can manifest. They exalt themselves and through the use of a patronizing attitude, they make their tormentors feel chagrined, thereby lessening the pressure. Thus, they sacrifice their affections for the favored few who "know."

The B2 myope always retreats to the world as he sees it, regardless of what is there. He is imprisoned behind his pride and his arrogant reserved exterior and suffers from isolation also, since communion is difficult when one fears to extend the self at the sacrifice of an arrogant pride and an assumed dignity and no consideration for any point of view except one's own. The B2 myope is not so adroit at manipulating persons and situations as is the B1 myope, so unless he can operate in a position of authority, such as director or executive, he is at a loss to find a niche. He is usually alert and guarded in manner and meticulous as to appearance with hair well groomed and clothes more formal. If sports clothes are worn, they are not casually displayed but are clean, neat and worn well. He keeps his surroundings to perfection also. As with the B1 myope, the degree of myopia reveals the level to which these attitudes are ingrained in his psyche.

Nichols offers a theoretical basis for the origin of myopia that is quite different from the traditional Skeffington model. To Nichols, myopia starts in the cradle--not at school age or when the child starts to read. It is not due to long hours of work within a restricted area of movement. At least that is not the fundamental cause of myopia. Every myopic child has either had experience or illness that has made him willful and prideful, and therefore insecure. His parents did not understand and tried to compensate by pleasing him to the exclusion of appropriate discipline, exalting his pride; or he was disciplined in a manner that indicates that his parents' will has taken precedence over the child's initiative, making him willful. According to Nichols, that is what myopia is. It is a maladjustment and an impressionistic approach to his situation according to his temperament. Another child, who is a realist, may make an entirely different approach to the same situation and become a strabismic. The parent attempts to substitute his own ideas about everything for the child's ideas about anything. One child, being eager to please, conforms. The other child, being determined and proud by nature, says to himself: "Listen here, I'm going to do as I please at whatever cost." The first child who wants to please, to conform, becomes a B1 myope. The child who will not conform, who is proud and arrogant, becomes the B2 myope. That is how temperament influences the case type.

The B1 Hyperope

As a temperament, the B1 hyperope operates primarily at the level of sensation. He has very sensitive antennae, being very sensitive to sounds, colors, texture, form and tone. He is sympathetic when he knows how you feel. He is curious about his world on a sensory basis. He likes to pick things up and feel them. He looks for a method, instructions, a way toward the goal, toward the objective (the HOW dimension). When the approach is appropriate, he is very sympathetic as a temperament and very delicate because he has very sensitive antennae. He is a promoter. He becomes a case type when his approach to sensation is blocked. He does not like to be told what to do. He loves to tell you. He tries to get what he wants through promotion. He gives so he can get something in return.

The B1 hyperope has two dimensions of the cube fairly well developed: the plane parallel to the self and perpendicular to the horizon and the plane parallel to the horizon. He knows himself and understands the world but does not know how to get where he is going without "killing everybody and himself in the rush" since he has only the emotional flow in through the subconscious and not the reasoning flow out of the conscious mind. This individual wants the center of attention at all times and therefore makes an audience of everyone and everybody he contacts.

As a child, he was forever asking questions but instead of waiting for an answer he would immediately go to the next one, trying to impress Mother. In this manner, he gets satisfaction but not achievement. He is a clever and brilliant child, of good appearance and charming so that everybody wants to do everything for him. The emotional flow is thus firmly established, with the reasoning process almost entirely lacking. There is no demand for thought, only for action through the emotional process toward the goal.

The B1 hyperope wants to love and be loved and to cooperate with others but he is not always equal to the impact of the confusion he encounters and therefore, he diverts his attention from the surrounds, as they exist, to a reflection of his own impressions often tuning out the negative reactions of those in his surrounds responsible for his discipline in early childhood.

In talking to a B1 hyperope, remember that he likes to hear about how he impresses the other fellow and how the other fellow feels about him. He loves the limelight and would like to know how he is doing with his world.

The B2 Hyperope

As a temperament, the B2 hyperope is active. He proceeds through action. He is a highly reasonable individual who functions primarily at the level of reason as a temperament. He is very sincere and works hard at having people understand his reasoning and his way of looking at people and things. He knows HOW to do things and does them well. He endeavors, however, to learn WHY things are true. As a child he is the one who asks the WHY questions. As a temperament, he is most appreciative and gracious. He knows the way and likes to share it. He loves to instruct and makes a good teacher. He learns as he teaches.

He becomes a case type when he asks "Why" but does not proceed through the process to learn WHY; or when his own children begin to ask "why" questions instead of "how" questions. As a case type, he scatters his energies instead of directing them toward the goal. He tends to become a jack-of-all-trades and master of none. He is reasonable, courteous, seldom rude but hard on himself and others, especially himself.

As a student, he is the nonachiever in school. He avoids the process of learning to read. He tries to avoid the process of finding out WHY. Being intellectually inclined, being highly reasonable and asking WHY without proceeding through the process to learn WHY, he becomes very certain that the answer to WHY is correct. Inasmuch as he has not earned the answer and has not proceeded through the process to obtain it, he is quite certain that whatever sounds logical, which may not be true, is correct. There is no one quite so "certain" as those who have not proceeded through the process in order to earn the truth. They become excellent critics.

The B2 hyperope sees everybody on his horizon but he has never known himself, for the emotional plane (the plane parallel to the self and perpendicular to the horizon) is completely blocked. No matter what you say to this person, he will love you anyway for he would try to reason out why you said what you did. He would probably think it out like this: "He's in some kind of trouble and has to pick on someone so it might as well be me."

He is always more interested in other people's achievement than his own and takes himself for granted. He does not depreciate himself too much, depending on how much he has achieved that is satisfactory to him. He thinks he must be all right for other people greet him in a friendly way, but he looks at the rest of the world with searching eyes, fascinated by the gyrations of others. This person knows that others are working their way through life too, so while he is compassionate, he is not too sympathetic for he thinks that it is a hard job for all of us and that "right is right" after all.

As a child he is taught to be considerate of others, that the highest virtue is to serve. He is taught that he is not always right and therefore should seldom form opinions, that things are seldom what they seem and he should never trust his impressions. Therefore, the B2 hyperope keeps things and people under observation and although he is not usually very hard on the other fellow, he is very hard on himself for he has high expectations for himself and does not form opinions about his fellow man except where he feels his rights are interfered with, for he would not be so presuming unless he felt his security was at stake.

When talking to the B2 hyperope, you can talk about anything on earth and it would not make any difference for he loves the world and the people in it. However, if you approach him saying, "Now, listen here. I'm going to tell you about you," since this is the realm in which he is least familiar, it is the last thing in the world he would want to hear about for you would then be talking in a dimension he lacks. He is not interested in hearing about himself, his traits, or being asked what he likes or does not like about himself.

The Self And The World

Nichols also briefly outlines how different people respond subconsciously to themselves to the hypothetical question that permeates every encounter: "Is the problem in the surrounds or is it in me?"

The emmetrope and hyperope are decisive and realistic. Their response is that "the problem is in the situation as it appears to me."

The myope is decisive and impressionistic. The B1 myope responds that "the situation is too great for me as I feel it." The B2 myope responds that "I am too great for the situation as I see it."

The astigmat is indecisive. He cannot make up his mind where the problem is. He just does not know whether it is in himself or in his world. He responds to the question by asking in turn, "Is the situation too great for me or am I too small for it?" and/or "Am I too great for the situation or is the situation not worthy of me?"

In summary, we have seen that, to Nichols, every person relates to his space world in a distinct way. The four major case types discussed are only minimal classifications. According to the Nichols model, "there are at least three or more different kinds of B1 myopes, several kinds of B1 hyperopes, several kinds of B2 hyperopes and several kinds of B2 myopes so that there are actually about twelve case types rather than four." What has been presented here, therefore, is only a small part of the Nichols approach.

What is significant about the Nichols approach is that she expounded the view that reflects personality, attitudes, behavioral and mental states at a time when most behavioral optometrists saw mental states as significantly reflecting visual status. Of even more importance is that she presented these ideas not one, five or ten years ago--but as far back as forty years ago.

Chapter 27

From Diagnosis to Therapy

Lenses and prisms are important elements in visual diagnosis and therapy. They change the value and quality of the light gradient as it enters the organism (Sutton 1975). A plus lens spreads the light scatter and decreases the light density. A minus lens constricts the light scatter and increases the light density. Prisms shift the direction of the incoming light.

What is less understandable is how and why the organism responds as it usually does to their effects, physiologically, psychologically and perceptually. It is traditionally accepted that convex lenses inhibit accommodation while concave lenses stimulate it. If so, how does the organism (or the visual system) "know" what to do? The fact that different people can demonstrate different quantitative and qualitative responses indicates that the response may not be as automatic and definitive as it is usually considered to be. What is important to understand is that the action of a lens or prism is on light and not on the organism. If the organism responds in given ways, it is only because the change in the nature of the light as it enters the eye mimics or simulates previously experienced natural changes in light stimulation (Forrest 1960d). Reactions to lenses and prisms are not reflex responses but reflections of responses that were first learned in a natural environment.

Lenses change the incoming light distribution from an object as if the object were either closer or further in space than it actually is. They simulate a z axis visual spatial change. Convex (plus) lenses alter the light distribution as if the object were further away in space. Concave (minus) lenses alter the light distribution as if the object were closer in space.

Prisms shift the direction of light on the x-y plane. When binocular base-out or base-in prisms are used, the dissimilar shifts in the direction of light entering each eye also simulate a z axis visual spatial change with only a minimal change in the light distribution gradient. With prisms there is a shift in the position rather than in the intensity or the quality of the light scatter on the retina. Binocular base-out prism changes the positioning of entering light as if the object were closer in space. Binocular base-in prism changes the positioning of entering light as if the object were further away in space.

Lenses and prisms, therefore, provoke responses that are directly related to the quality of previously learned experiences in focusing and converging for real objects in all different areas of real space and the manner in which those experiences were matched with touch, body movement and emotional feeling-tone. Through their ability to simulate elements of more natural events, lenses and prisms can be used as probes to determine how one has learned to handle different aspects of visual space; what spatial distortions, limitations, suppressions and concessions have occurred as a result of adaptation to specific stressor agents as well as to the stress that we ourselves create by our attitudes and emotional responses.

Diagnostic Relationships

A standard convergence push-up or nearpoint of convergence test is a measure of the ability to follow visually a target as it moves inward toward the eyes until fusion can no longer be maintained. In this procedure, the target actually moves in space.

Prism vergence (duction) tests measure the capability and flexibility to align closer and further and for the eyes to move in and out while one observes an object that is stationary in space.

Getman (1977a) prefers the use of the term ductions rather than vergences. He makes the distinction that "vergences are the movements of eyes (while) ductions are the measurements of interpretations of changes in spatial locations and orientations." We have preferred to use the term vergence since it is more universally accepted while carefully differentiating when it refers to the mechanics of the test and when it relates to perception and performance.

The first element in the standard vergence test is the point of blur. This is a measure of the flexibility in the system or how much convergence and divergence can be altered as base-out and then base-in prism is slowly increased before accommodation is affected. Getman (1977b) considers this as a test of "the ranges of freedom in comprehension (that operate) in spite of stresses from any distortion of the visual whole."

The second element is the "break" finding. This is a measure of the degree to which the convergence and divergence aspects of the centering process can be manipulated before double vision (diplopia) or suppression of one visual circuit occurs. At one level it reveals how far in and out in space binocular centering can be manipulated while fixating a stationary and stable target before fusion is lost. At another level, it reveals the limits of play within the somatic (skeletal) system before the system gives up.

To Getman (1977b), the break findings are "a measure of the ranges of freedom in the visual spatial whole that will permit the maintenance of orientation in spite of the stresses from any distortion in this visual whole."

The third element of the vergence test is the "recovery." This finding represents the ability of the vergence system to recoup or reorganize after binocular function has been disrupted. As Getman puts it, the recovery findings "are a measure of ranges of survival--how quickly and satisfactorily the patient can get his disrupted world back together again."

All these vergence tests measure spatial operation using the plane of regard as a locus. As emphasized before, changes in centering are triggered by the stimulus being altered to simulate spatial shifts while actually remaining in the same position. The fusional ranges and binocular function that are probed by these tests should be relatively balanced in front of and behind the fixation point, at least at near. Since one does not usually look through a solid distance object or gaze beyond the horizon, the divergence measurement at distance is normally expected to be lower than the corresponding distance convergence finding. Distortions found in these measurements reflect and represent warps in the perceiving and handling of natural visual space.

If all vergence break findings are low, it is usually representative of an overly tense, high-strung individual. It can also reflect a general systemic problem. A significantly and persistently low divergence finding at distance (#11) usually indicates an individual who is internally verbal and tense and has a decreased, often repressed, ability to image, fantasize and to let go of embedded concepts (Forrest 1981c).

Morse-Peckham (1926) noted that "if there is freedom from stress, ductions are high. If there is tension, and if there is obstruction to the delivery of commands from the central nervous system, ductions are low." Skeffington (1961) used the following guideline: "When the blur is low the case is new. When the break is low the case is old. When recoveries are low adaptation is poor."

Prism phoria tests also measure spatial orientation using the fixation plane as a locus. To Skeffington (1957b) the phoria provides information "as to where the person being examined is operating along the range of his centering, in the visual space world that he himself has structured." It reveals how he "is computing space and what prior experiences are summoned by the visual patterning" (Skeffington 1965b). Getman (1978a) notes that if judgments in visual space "are consistent, the phoria and duction findings will be consistent. If the individual is insecure and uncertain about his spatial judgments of location and orientation, his phoria and duction findings will be variable and uncertain." The Stress Phoria reveals the conditioned spatial computing responses to visual stress (see chapter 22) (Forrest 1960c).

Lens tests as well as prism tests probe visual spatial operation. In fact, the entire optometric evaluation sequence (called the "analytical examination" by A. M. Skeffington [1957g]) "is a series of measurements yielding information as to how the patient is operating in his visual space world." Skeffington (1957f) strongly felt that "the connective 'and' must be eliminated from the thinking of space and person. 'Space-person' indicates that space does not exist without the person, nor the person without the space." This relationship between the person and his space world always enters into the adaptive outcome that is "chosen." The individual decides for emmetropia. "He may choose the constricted space volume of the myope. He may elect asymmetry of the total ground as an astigmat, or of the total space world as an anisometrope. The interplay between the person and the environment decides what will be done" (Skeffington 1965c). Since the operation of the autonomic and skeletal nervous systems as they are reflected in the visual process controls the interpretation of visual space, the analytical examination also serves to probe the relationship between these systems and their corresponding effects on an individual's personal space world. In general, near tests reveal the primary actions and reactions to visual stress while distance tests reflect the spatial rigidity and pervasiveness of the visual stress response and all that it represents.

Diagnostic information can be interpreted in various ways. Skeffington, for example, conceived of vision as being a spatial event and he viewed the analytical examination as probing how an individual structures his space world. Harmon (1958,1966b) conceived of visual spatial action as being predominantly a reflection of how one's postural status and internal coordinate system relates to gravity and he perceived the analytical examination as a means of probing the end results of postural status as it might be reflected in the visual system.

Those who hold a more physiologically oriented model of the visual process may accept the spatial and postural attributes of vision but feel more comfortable with the reflection of these actions in terms of accommodation and convergence mechanisms. For them, the analytical examination reveals what is happening within the accommodative-convergence systems as well as in the neural organization of these processes.

From the point of view that consciousness is the primary element in this interaction, the analytical examination is seen to reveal not only spatial, postural and accommodative-convergence elements but also the end result of one's biases, attitudes, self-image and basic belief systems through which one's space world and one's actions in space are structured (Forrest 1984d).

Vision Therapy

There are three major categories of vision therapy: palliative, remedial and preventive. Palliative vision therapy consists of the use of visual aids to alleviate the signs and symptoms of structured ocular conditions, including decreased visual acuity, rather than to reduce or eliminate the actual problem. This type of therapy is helpful in ocular difficulties of obvious genetic origin, in cases of disease or injury, and where neglected functional problems have resulted in permanent changes in ocular structure. Remedial therapy is geared toward enhancing visual function, reducing and negating the effects of stress and making the individual a better performer on all tasks dependent on visual efficiency. Preventive therapy is the use of visual aids and guidance protectively, to try to prevent the onset of a visual problem.

Some of the behavioral tools used in remedying, preventing and palliating visual problems are: (1) all forms of lenses, (2) all combinations of prisms, (3) all techniques and instruments utilized as part of a visual training program, and (4) recommendations as to reducing, removing or avoiding stress- producing factors such as improved diet, better lighting conditions, more appropriate seating arrangements, enhanced work conditions, better posture and the use of relaxation procedures.

There are different rationales for using these various therapeutic tools. Some view lenses, for example, as reducing accommodative demand or accommodative effort. Skeffington (1957e) saw the plus lens as "a large and very essential factor in freeing the organism from some measure of absorption of the vital degrees of freedom to meet the demands of the culture." Getman (1978c) feels that "the power of the lens is not in its dioptrics and focal lengths (but in its) influence on the interpretative processes of the visual system, and its relationships to all other information systems....The lens allows reduction and elimination of confusions and delays. Lenses influence all information systems by influencing the light signals that reach the visual system through the eyes."

Sutton (1975) notes that a plus lens can be used to emphasize ground and deemphasize figure while a minus lens can be used to emphasize figure and deemphasize ground. Kraskin (1981-83) views convex (plus) lenses as reducing the tonicity of the posturing muscles of the body while concave (minus) lenses are seen to increase postural muscle tonicity. Larrabee and Jones (1980), exploring the relationship between low-plus lenses and near performance, found that judiciously prescribed low-plus lenses resulted in significant behavioral changes on paper-and-pencil tasks.

Kaplan (1979) relates the prescribing of bases-up and bases-down yoked vertical prisms to a number of factors including the individual's preferred central-peripheral (eso or exo) processing style, the accommodative-convergence relationship and whether or not specific visual-motor and peripheral awareness performances are improved through the use of the appropriate yoked prisms. In general, bases-up yoked prisms are used for the more peripheral (exo) individuals and bases-down yoked prisms for the more central (eso) types.

Kraskin (1981-83) views yoked prisms as influencing the process of orientation. Yoked prisms cause the eyes to move. Eye movements result in a shift of the pelvis, causing a corresponding shift in the position of the center of gravity. The Kraskin approach in using yoked prisms therapeutically is to induce a stress in the visual system in the direction of the existing asymmetry in the hope that "the system will respond to reduce the stress and, thus, alter the asymmetry in the desired direction." Performance tests are used as guides to determine whether the patient will benefit from the use of yoked prisms. If specific directional yoked prisms are to be effective, they should cause a decrement in performance on standard polaroid stereopsis tests.

The Role of Visual Training

Visual training is a term that encompasses varying approaches and points of view. To many, it involves improving muscle strength, agility and flexibility. To others, it is a matter of improving the efficiency of basic visual abilities such as oculomotor skills, binocularity, accommodative-convergence relationship, hand-eye coordination, visual-postural-gross motor organization, intersensory-motor integration, visualization and visual memory function.

Some practitioners stress instrument training and others stress training only in unenclosed space. Some stress higher order visual skills while others stress gross motor supportive abilities. Some stress the prescribing of lenses and others minimize their use until the therapy is completed. Some emphasize awareness and visualization skills while others stress hand-eye coordination and fusion tasks. It is intriguing how so many different procedures and approaches can accomplish the same goal and how many techniques that work effectively for one practitioner are ineffective when used by another. Obviously there are common elements among all the seemingly diverse approaches.

Behaviorally, training of any kind is a controlled and repetitive procedure that follows he laws of learning. It consists in presenting, more or less sequentially, the conditions for learning new skills. Regardless of what devices are utilized, whether they be lenses, prisms, stereoscopic-type instruments, rotary devices, polaroids, anaglyphs or balance boards, the desired goal is to change the behavior of the individual by setting up different learning demands. Visual training is not limited to eyes. It aims at unifying visual functioning with total body action through a consistent matching of all sensorimotor modalities. This results in improved visual performance and an enlarged and more flexible operation of the individual within his visual space world.

Skeffington (1957a) realized that stressor agents had less impact on vision when there were greater degrees of freedom within the process so he saw visual training as a methodology for arranging conditions "for learning additional degrees of freedom to absorb stress." Training procedures were viewed as being methods for stimulating one mechanism more than other mechanisms so as to build latitudes or degrees of freedom between them (Skeffington 1953,1970). He saw other goals of visual training as the establishing of "good, effective, unimpeded output in all the measurable and observable visual behaviors" and the developing or evoking of "good, solid, full space awareness" (Skeffington 1958b).

Getman (1978b) sees visual training as being "the opportunity given the patient to build integrations of all his interpretative systems with the visual system as the guide and monitor for the accuracy and effectivity of all systems in all decision-making processes." He also stresses that "for vision training routines to belong in an optometric regimen and be relevant to optometric purposes and responsibilities the procedures used should involve visually directed, visually appraised, visually monitored, and visually corrected actions."

Macdonald (1965) lists the purposes of visual training to be: (1) getting "the eyes to lead the body comfortably, easily and effectively," (2) "to develop within the body system the degrees of freedom necessary for the organism to adjust and adapt to the various asymmetries within the visual task," and (3) "to develop an awareness of the part the visual process plays in the process of thinking." He also considered a goal of visual training as being "the establishment of a balanced, dynamic ratio of trees (specifics) to forest (generalities)" or the ability to see space as a solid and to be able to direct attention to any area within the solid while maintaining awareness of the solid (Macdonald 1962).

Visual Training and Processing Style

A model for understanding adaptive outcomes in terms of this type of central-peripheral (figure-ground) interaction was discussed in chapter 23. Knowing that the optimum role of efficient central-peripheral organization is to be able to localize visually and fixate some area of space while maintaining a full awareness of the surrounding environment provides some direction for therapy (Forrest 1976b).

Therapy for the "either-or" exo processor involves the simultaneous unification of central and peripheral information. The exo processor needs to learn to draw meaning from the center and the periphery at the same time. Bitemporal strips (one-eighth-inch black strips or rubber bands placed on eyeglasses temporal to the pupils) help create awareness of both center and periphery simultaneously, provided the patient is encouraged to look straight ahead at something while being conscious of the strips on the eyeglass lenses. Drawing bilateral circles on a chalkboard can also be a valuable training procedure provided that a meaningful target such as a square of small letters which is to be read aloud is fixated centrally while the bilateral circles are being drawn.

Training for the "central" eso processor entails enhancing peripheral meaning-getting ability. Only when that aspect becomes efficient should it be integrated with central activities. Binasal patches are helpful in these cases with the patient being encouraged to look straight ahead toward or into the patches while being aware of the periphery. In this way it helps increase peripheral sensitivity in the eso processor who is weakest in this area. Drawing bilateral chalkboard circles can also be useful provided it is done with a meaningless central fixation target such as a simple X with the emphasis placed on peripheral awareness.

Birnbaum (1978) relates the exo-processing style to a right hemispheric (spatial, simultaneous, global, field-dependent) mode of consciousness and the eso-processing style to a left hemispheric (analytical, sequential, linear, field-independent) preference. In training, he recommends procedures for the exo processor that foster left hemispheric function with an emphasis placed on central, analytical and sequential type tasks. For eso processors,the emphasis is placed on enhancing right hemispheric function through the use of peripheral awareness, imagery and passive contemplation.

In a different attempt to reach the same goal, Fritz and Fehmi (1982, Fehmi & Fritz 1980) have developed a series of activities they call Open Focus training. Open Focus has as its goal the broadening of awareness "to simultaneously include all those perceptible events which are salient in the nervous system." In reading, their aim is to foster reading effortlessly while attention is gradually expanded to include thoughts, emotions, peripheral visual experiences, tastes, smells and sounds. As part of their training program they ask: "Is it possible for you to imagine the space between your eyes...the space between your ears...that your whole head and face are simultaneously filled with space...that your whole being fills with air when you inhale and your whole being is left filled with space when you exhale...that space inside and outside become one continuous and unified space...that all your experience is permeated and pervaded by space?"

Feedback In Visual Training

Each and every interaction has some effect on behavior and modifies it to some degree. No two exposures to the same situation mean exactly the same to the organism nor do they have exactly the same effect. One always comes to a second encounter with much more than was brought to the first. An individual, therefore, is not quite the same person as time progresses.

The fact that behavior changes suggests, in most instances, that some form of learning has occurred. Behavior modification occurs simply from existing, from one's interaction with the environment. It can also be intentionally and formally guided by outside forces such as is the function of education. It can also be self-directed. Whatever form it takes, it involves a number of operational factors, many of which go by the name of feedback or biofeedback (Forrest 1981b).

As was noted in the section on biofeedback in chapter 16, feedback has become a very popular term nowadays. It represents far more than what the original cyberneticists meant when the word was first coined. Today, it can refer to the operation of a single neural circuiting, a multisystem circuiting in the way that one sensorimotor system can be used to be aware of what another is doing, or it can represent just a simple conscious awareness of a particular body function. It can refer to both a conscious or subconscious activity. Its traditional reference was to an act of monitoring (Weiner 1949). Its popular reference is to an act of control, especially over biological processes that were thought to be somewhat uncontrollable. In this sense, feedback or biofeedback refers to the conscious awareness of an act and the use of this awareness to modify the act. With some internal behaviors it has become advantageous to use electronic devices that amplify the response and allow it to be consciously monitored visually through the use of lights and oscilloscopes or aurally through the use of varying types of sounds or sound intensities.

It has been used for relaxation (Brown 1977), migraine (Koppman, McDonald & Kunzel 1974; Green & Green 1977) and cardiovascular problems (Brown 1977).

A more important element in the feedback process is that all the varied methods of feedback monitoring are built upon a superstructure in which the central mechanisms of the brain constantly bias the system. This biasing is called feedforward and is often mediated by gamma-efferent fibers which terminate in specialized sensory receptors rather than in muscle tissue (Pribram 1971). Feedforward is the method by which all actions are influenced and guided by the preferences, commitments and attitudes of the organism.

Although biofeedback has been labeled as a new type of therapy, to optometrists involved in visual training its principles are far from new. In visual training, the information patients receive from the use of polaroids, anaglyphs, lenses and prisms provides them with the knowledge as to what is occurring and what they might have to do to alter the existing status. In fact, as Birnbaum (1984a) states, "the most powerful vision therapy techniques are those which provide effective feedback to the patient."

Ciuffreda (1982) has successfully used auditory feedback to control nystagmus. Flom (1980) has used auditory feedback to control eccentric fixation in amblyopic eyes. Biofeedback has also been used to reduce the degree of strabismus (Hirons et al.1978; Van Brocklin et al.1981; Goldrich 1983). Balliet, Clay and Blood (1982) were able to improve visual acuity levels in myopes using a computerized optometer with the help of buzzers to monitor whether the response was correct or incorrect. Trachtman (1978) also has successfully demonstrated improved visual acuity in myopes using an infrared optometer with auditory feedback. He notes that although visual acuity improves first, over time the myopia itself is found to reduce (Trachtman 1984; Trachtman, Giambalvo & Feldman 1981).

The Control of Myopia

Many of those involved with vision therapy have been interested in the idea that refractive states, especially myopia, can be stabilized or reversed. In most instances, the assumption is that functional myopia is primarily accommodative in nature. A number of studies have demonstrated that myopia can be controlled through the use of atropine to inhibit accommodation (Young 1965b; Bedrossian 1971; Gimbel 1973; Kelly, Chatfield & Tustin 1975). Others have indicated a similar positive effect in controlling myopia through the use of appropriately prescribed near-vision lenses or bifocals (Parker 1958, Roberts & Banford 1967,

Ludlam 1968, Oakley & Young 1975, Greenspan 1981). Although Trachtman (1978,1984; Trachtman, Giambalvo & Feldman 1981) has reported changes in the status of the myopia itself, many research studies have indicated that visual training for myopia more often results in an improvement of visual acuity with the refractive status remaining relatively unchanged (Marg 1952, Hildreth 1947, Woods 1946).

Ray Morse-Peckham (1939) noted that German ophthalmologists between 1850 and 1900 routinely prescribed base-in prisms for all myopes to restrain the progression of the condition. Base-in prisms were seen as stimulating negative vergence which hopefully would be accompanied by negative accommodation which, in turn, would reduce the myopia and the minus lens power needed for clear distance vision. For incipient cases of myopia, Morse-Peckham would start treatment by prescribing bifocals with Plano base-in prisms (one diopter for each eye) for distance with a plus sphere add for near.

Birnbaum (1979,1981a,1981b) differentiates among three types of myopes: the incipient myope, the progressing myope, and the stable myope. Recommended therapy for the incipient and progressive myope consists of: (1) prescribing appropriate plus lenses for near to reduce the accommodative demand, relieve accommodative insufficiency, and reduce the accommodatively induced overconvergence tendency, (2) a visual training program to improve accommodative efficiency and (3) advice concerning visual hygiene to reduce stress on the visual system.

The visual hygiene instructions consist of not holding reading material too close, use of good lighting, taking short (one-to- two minute) rest periods when doing close work, looking across the room at the end of a page of reading and maintaining general awareness of the surrounds when reading to encourage "a less intense, more relaxed attitudinal set."

Birnbaum feels that once myopia has stabilized and its progression ended that the use of plus lenses at near is no longer indicated and that visual training with the aim of reducing or eliminating the myopia is generally unsuccessful.

Friedman (1981) recommends varied techniques utilizing lenses, prisms and visual training devices in order to teach patients to inhibit and stimulate accommodation efficiently and flexibly, to eliminate suppressions, to enhance binocularity and to induce relaxation, visual imagery and an awareness of periphery. One procedure he uses to encourage awareness and control of the accommodative process requires the patient to clear a high-minus lens and then blur the vision through low-power lenses by trying not to concentrate. When attempting to stimulate accommodation ("on-accommodation") the instructions are to "look close" and "concentrate hard." When attempting to inhibit accommodation ("off-accommodation") the instructions are to achieve a state of maximum relaxation.

Friedman also stresses the observance of good visual habits both for the management and prevention of myopia. He particularly recommends avoiding excessive visual concentration, especially involving scrutinizing details; periodically (at least five to ten times a day) becoming generally aware of peripheral visual space; avoiding staring where the eyes are held rigidly; avoiding straining to see when viewing conditions are not optimal (e.g., low illumination, glare, poor visibility, or when fatigued); avoiding straining to see by broadening one's visual field by passively observing all the changing aspects going on in the periphery; avoiding being in situations that can provoke straining to see; and periodically looking far away either in natural space or by closing the eyes and visualizing a relaxing outdoor, distant scene.

The Control of Hyperopia

Very few professionals have been as concerned about moderate or high hyperopia as they have been about myopia. Trachtman, for one, using the infrared optometer with auditory feedback, has reported some success in controlling and reversing hyperopia. Miele (1985) has also noted significant results in reducing hyperopia through the use of a minus lens procedure.

In the Miele approach, patients are screened through the use of plus and minus subjective tests. The plus subjective test is a measure of the maximum amount of plus through which the patient can comfortably read 20/20 on the distance Snellen chart. The finding is obtained by starting with the plus found on the near (twenty inches) dynamic retinoscopy test (#5) and slowly reducing the power until the 20/20 line can be read. The patient is considered to be a candidate for therapy if he is found to have 1.25 diopters or more of hyperopia.

The minus subjective tests consists of placing -4.00 diopter lenses before the patient and slowly reducing the power until the maximum minus is reached through which the distance 20/20 line can be read. The maximum plus and minus found on these two subjective tests determine the limits of a "range of clarity" at distance.

The most important aspect of the therapeutic program to reduce hyperopia is in the use of minus "retainer" lenses. These are prescribed either as single-vision lenses or in the form of a bifocal with the minus power in the distance portion. For the minus to be effective, some degree of minus power has to be demonstrated in the "range of clarity" measurement. If none is found due to the minus subjective test being in plus, then reduced-plus retainer lenses can be temporarily prescribed in an attempt to prod the "range of clarity" into shifting so that some minus becomes acceptable.

The tolerance of the patient to the minus power determines when the retainer lenses are worn. Sensitive individuals may be instructed to wear them only for television and for at least five minutes before starting to read. Others may be able to wear them for more extended periods of time.

A regular full plus prescription can be used for critical seeing until refractive changes are noted. If, over a period of a few months, visual acuity and comfort improve, the plus in the basic prescription can be reduced while the patient continues to wear the minus retainer lenses.

Chapter 28

A Psycho-Behavioral Philosophy

One aim of this book has been to survey major concepts in the area of stress and visual stress and the premises on which they are based. Actually, there is no such thing as an objective summary or review of the literature. The process of choosing what to include, what to leave out, what to emphasize and how it should be emphasized always reflects the prejudices of the writer. This book is no exception. It has a definite bias. It openly supports what has been called a psycho-behavioral approach to the subject--the assertion that consciousness in its many forms underlies all aspects of behavior. Many elements of this philosophy are incorporated throughout the book. Some have not been touched upon.

Innate Factors

One aspect that has not been discussed so far relates to the importance of innate factors (Forrest 1984d,1984a). It is interesting how we tend to accept the concept that human beings are born with certain physiological givens such as basic sensorimotor reflexes and autonomic functioning (respiration, circulation, digestion, elimination, etc.) while often believing that behavioral states (attitudes, emotions, personality, style, etc.) as well as one's knowledge about space and the things in space, including one's self, all start from absolute zero; that the mind, as John Locke (1965) put it, is a tabula rasa, a blank slate at birth and that psychological and behavioral factors are established mainly through experience.

It would seem reasonable that what is good for the physiological goose is good for the psychological gander. It would also appear logical that since we are spatial organisms, operating in space and knowing something about the space around us that one's content of consciousness, including psychological state, spatial ability and visual imagery ability, should start at birth with some degree of preprogrammed, encoded knowledge. It is difficult to conceive that all or most of what we are is determined from birth-trauma on or possibly from some prenatal form of experience. Much more appears to be built into the perceptual, cognitive and affective domains than many of our current premises suggest. One indication of this is seen in the newborn's basic repertoire of facial expressions which appears to represent appropriately what the infant feels at the moment (Trotter 1983).

There is a growing body of evidence to indicate that we are all born with a basic psychological or "mental states" intelligence (a psycho-"logic") which, though primitive, consists of a rudimentary spatial map, primordial visual imagery ability, and primary emotions. Onto this inborn repertoire are superimposed early learned interactions and belief systems which then influence one's attitudes and motivations toward people, objects and events. This, in turn, casts an influence on both the individual's operational style which includes his perceptual, cognitive, personality and visual styles and his basic biological intelligence (bio-"logic") affecting physiological function in the autonomic (visceral) and somatic (skeletal) domains. It is this latter aspect which is seen to influence accommodative- convergence relationships and postural status. Although sensorimotor experiences and visual-postural relationships can and do strongly affect one's personality, attitudes and emotional state, they function only on the matrix of an already operating, inborn, psychological and behavioral base.

The "In-Out" Factor

A second aspect of this psycho-behavioral philosophy pertains to figure-ground interaction or central-peripheral organization. The relationship of figure to ground or center to periphery are elements of an even more basic expression, the fact that all existence appears to reflect "in-out" characteristics that operate from hypothetical centers.

This is true of all physical phenomena where electrons with their surrounding charges, planets, moons, suns or galaxies are either in or out in relation to a center.

Sidedness may be inherent in symmetrical organisms or objects but sides always relate to a center. They are always out in relation to something in. The labels right and left are simply arbitrary cultural constructs. Rightness and leftness do not appear to exist in nature. In-out relationships do.

Up and down appear to be obvious physical directional attributes. We learn that what is closest to the ground is down and what is furthest away is up. When we stand, our feet are down and our heads are up. If we lie prone on our back, our back is down and the front of our body is up. Yet, this too is an illusion, a man- made conclusion. If we picture the Earth as it is in space and then visualize people standing at the north pole, south pole, equator and elsewhere, it becomes obvious that the effects of the force of gravity are also an in-out phenomenon. People, things and their parts are either in or out in relation to the center of the Earth. Otherwise, if people are standing at the north and south poles, someone must be upside down.

Biological function also operates from hypothetical interacting centers. There is a head "center," a total body "center" (center of gravity), an ocular "center" (fovea) and a binocular "center" (cyclopean eye). Some even consider that there is a motor center and an emotional center.

Consciousness exhibits in-out relationships as well. Its externalized center is the ego. From this ego-center all people, objects and events are interpreted as to how we, as center, relate to it (out there) and how we, as center, relate it to it from our individual points of view. It is from this egocentric locus that we construct our perceptual, cognitive and visual space worlds. The functional in-out (figure-ground, central-peripheral) behaviors we develop stem from learned biases, belief systems and attitudes. Even the intuitive level of mind can be considered to have a center which can be called, for lack of a better name, an inner ego.

The Role of Perception

A third element in this approach concerns a somewhat less conventional perspective of the role of perception. In this view, all perception is seen as a limitation, as a filter superimposed over what "is." It is the method we use to construct a reality for ourselves. When we perceive, we literally conceal aspects of reality. We camouflage it. We interpret it. "Our state of mind...determines our perceptions....Our inner world, our level of consciousness...selectively creates the world we perceive....That which affects our mind affects our perception" (Harvey 1983).

Ordinary perception (not intuition or ESP) is primarily dependent on information from the senses. Since the senses provide only selective information, perception at its best has to be a limited outlook. It may be all we have to work with but that fact should not stop us from recognizing what it is.

Perception is evaluation and judgment, and judgment of any kind is always divisive. Perception is selection. We perceive only parts within a whole, not the whole. Categorizing is an extension of this process. It represents the basic human tendency to group together those aspects we perceive as being similar. Other aspects are either not perceived at all, are rejected as being inconsequential or deemed negative and destructive and therefore to be avoided.

Perception is simply a way of looking at things with the colorations we bring to a situation based on our past experiences and attitudes. It is also greatly influenced by internal verbalization which reinforces these past attitudes. Every optometrist who has worked with variable three-dimensional polaroid vectograms has seen an example of this. As base-out prism effect is induced, the viewer normally perceives the resulting image as becoming smaller and closer. As base-in prismatic effect is induced, the image appears to get larger and further away. This is known as the SILO effect (smaller-in, larger-out). Some people perceive a SOLI (smaller-out, larger-in) effect. Many of those who exhibit a SOLI response are able to localize the position of the image more accurately with a pointer or their finger. It is interesting to note the response of these individuals when they are asked why their fingers seem to "see" things differently than their eyes alone. As they are encouraged to become more aware of what goes on in their minds as they observe,it soon becomes clear to them that the SOLI response represents the fact that what one expects overrides basic optics and physiology. It also becomes apparent that perception is not necessarily representative of reality. It is a construction, a choice, a selection. Perception is very easily influenced by what one looks for, by the influence of reason and by the effects of internal verbalization--all of which are bound to past experience. A SOLI response often occurs in an individual whose perception is more internally oriented, whose priority and trust lies with reasoning and past experience. It is more often found in individuals who are more rigid in their attitudes and who resist seeing things as they are in order to provide a sense of consistency and stability against the appearance of change.

Although this is one example of obviously aberrant perception, perception of any kind, being an act of interpretation, is never true. It varies from person to person and from time to time. It is always a mental creation in time and space. This means that even when we look at someone, we never totally see that person as he or she truly is now. We see that person in terms of our past interactions. We see that person in terms of learned biases based on skin color, sex, weight, language, physical attributes or dress. We see a constructed image rather than what is really stimulating our senses.

In addition, if our perception of the external world is basically an interpretation, can we ever be objective about ourselves? As with our external perceptions, more often than not, we are not able to know ourselves; we only "perceive" ourselves in certain ways. We draw on the material of past experience and selectively interpret ourselves or aspects of ourselves in positive and negative terms. By doing this, we develop an "image" of ourselves. Then, when we view or perceive the world, we do so as an "image viewing images;" as a construction viewing a construction; as an ego-bias viewing a stimulus-bias; as an illusion viewing an illusion. When we view the world, who then is the "see-er" and who or what is the "seen" when both the see- er and the seen, the observer and the observed are simply constructed images (Krishnamurti 1968)?

As abstract as this may be, it has a very real relationship to education in general and to vision therapy in particular. In therapy, is a patient directly or indirectly being induced, in the name of good perception to reinforce his existing images or biases about people, objects or events? Is he being taught that the therapist's illusions and images are the ones he should develop? Or, is he being encouraged to become more generally aware about what he observes and how he views others and himself? Can this latter goal, as ideal as it may be, be aided or directed by a therapist who views the trainee, life and himself through images? It is something to consider. It also encourages us to rethink what is really meant by terms such as imagery, visualization, visual training and, more importantly, "vision."

The Role of the Ego

A fourth element in this personal psycho-behavioral philosophy pertains to the role of the ego as a fundamental component of mental function (see chapter 15). The ego is a necessary mechanism for human life. It establishes psychological individuality and is a primary component of figure-ground extraction and attentional states. It is as important psychologically as the fovea is visually. Having a fovea does not automatically make one an overly central visual processor. Similarly, having an ego does not automatically mean that one has to become overly egocentric. It is how these functions are utilized that is important. An efficiently operating visual system requires a flexible action between central and peripheral information processing at all levels--both within the visual process and in the interaction of the visual process with other sensorimotor systems. An efficiently operating mental process requires an equally flexible interaction between ego function and the other levels of mind. The similarity being drawn between the functions of the ego and the fovea is not just a convenient analogy. Both are part of the same overall process. This is why a visual training program geared to improving visual and intersensory-motor central-peripheral interactions has the capability of having this improvement generalized to mental- states function. This is also why enhancing mental-states function and the relation of the ego to more intuitive mental levels can result in improvement in visual function. The process is bidirectional though, if priorities have to be assigned, it is the mental-states factor and all that it represents that appears to be primary.

Split-brain research has indicated different functions for the two cerebral hemispheres (Sperry 1964, Gazzaniga 1967, Ornstein 1977, Gazzaniga & LeDoux 1978). This has tempted many researchers to attribute various types of behaviors to the action of either the right or left cerebral hemisphere. Some have assigned ego functions to the left hemisphere and more intuitive functions to the right hemisphere. Since analytical language activities in most individuals is seen to be related to left hemisphere function, some have considered that emotion and imagery are related to right hemisphere action. This type of attribution has been intentionally avoided throughout this book for a few reasons. First, so little is known about hemispheric operation. Second, there is a great deal of knowledgeable opposition to most views presented on the subject. Third, there is still a strong divergence of opinion as to whether intuitive and ego states utilize brain mechanisms as a medium for translating thought into body action in three-dimensional space or whether mental states are derived solely from brain function.

The Stress Response

A fifth element pertains to the subject of stress and visual stress. The interpretive and judgmental aspects of ego function are seen to be the direct cause of stress. Stress is not in the event. Events are neutral. The major emotional and physiological overtones of any given interaction come solely from the individual and not the event. This means that visual stress does not primarily come from the act of reading or from other types of close work. Neither does persistent postural distortion come primarily from poor lighting or poor posture when doing near- vision tasks. Stress comes from one's approach to a task and the often subconscious interpretations and judgments that act as a pre-set. This affects how one functions by inducing muscular rigidity that can persist even when the task is completed. The negative effects of restraint, tension, immobilization, sensory underload and sensory overload all stem from the inability of the individual to flow with events rather than from physical or physiological stressor agents alone. The principal factors are those mental attitudes that are generated by deeply hidden feelings of inadequacy, vulnerability, helplessness and the fear of change.

The Primacy of Consciousness

A sixth premise is that the body, including the ocular mechanisms, reflects and becomes a symbol for the way consciousness expresses itself through beliefs and attitudes about life and life experiences. Except for pure reflex functions, the process of vision does not respond to a stimulus. Consciousness does. The process of vision and the ocular mechanisms are tools by which consciousness localizes and identifies. Vision is an emergent of an efficient use of consciousness. Consciousness is not an emergent of an efficiently operating visual process, although the efficiency of the visual process as it interacts with other body processes does have an enhancing effect on consciousness.

Skeffington (1963) proposed a "four-circles" paradigm to illustrate how four major human operational systems interact and how, from this interaction, something greater than the sum of its parts emerges. He called this emergent aspect VISION. Sutton (1985) lists the separate systems that each of the four circles represents and elaborates on what their actions signify to the organism. He sees the "Anti-Gravity-Orientation" system as answering the question, "Where am I in space and time?"; the "Centering-Localization" system as answering the question, "Where is it in space and time?"; the "Identification" system answering the question, "What is it?"; and the "Speech-Auditory- Communication" system as answering the question, "How do I communicate impressions of 'where' and 'what'?" What remains unstated, yet inferred, is that the most important factor behind each of these questions is not some aspect of the process of vision but consciousness itself. It is consciousness that is interested in and gets the "message" in each case.

Skeffington (1938) realized the importance of consciousness and mental states when he noted that "the only way we gain a concept is through mind....When mind grows uninterested in gaining information from a certain point, seeing becomes unimportant at that time."

Lovgren (1977) elaborates on the same thought when he states that "seeing out-of-focus is also thinking out-of-focus and imagining loosely....Focusing is truly a nonverbal act of thinking, taking a position, selecting a point of view." He continues by observing that "knowingly or not, we instruct our eye what to look for. Obediently it picks and chooses every moment of our lives, even while we are idling or daydreaming. Some sights are more important than others, but they all add up to what we select to be. Our inner seeing establishes policy and the outer eye obeys. This is one act, not two."

In a similar vein, Krishnamurti (1982) states that "the world outside is the expression of the world inside; there is no division between the outer and the inner." And John Milton (1608- 1674) postulated the same idea when he noted that "The mind is its own place, and in itself can make a heaven of Hell, a hell of Heaven."

From Beliefs to Outcomes

A seventh element in this philosophy holds that there is a loose behavioral hierarchy that leads from mental states to adaptive outcomes. Though there is much overlap and bidirectional movement within this hierarchy, there is still a strong unidirectional quality to its basic structure. Its stages include the establishment of early symbolic relationships, the selective recall of past experiences, the formation of core beliefs, perception, thought, imagery, affect (emotion), overt behavioral effects and finally, adaptive outcomes.

All human life from infancy on deals with symbols. Words, images and objects are all utilized as symbols. Piaget (1952) hypothesizes that during the early months of the sensorimotor stage, a child's knowledge of the world is based almost exclusively on basic sensorimotor experience. An object exists to an infant only when it is observed. If it is out of sight, it is out of mind. Piaget, however, does not seem to acknowledge the possibility that by matching sensorimotor experiences with emotional counterparts, a child learns that events can both trigger and represent given emotions and that, in turn, given emotional states can both trigger and represent the recollection of events. In other words, through basic sensorimotor-emotional interaction, it is possible for a child very early in life to learn to handle emotional symbols and to do so long before one external object or attribute becomes a symbol for another. In this view, affect has the potential for becoming a symbol for experience and through its use, a child can become involved in a form of symbolic thinking much sooner than Piaget (1971) seems to indicate (Forrest 1984d).

Over time, with an increased use of language and imagery, more sophisticated emotionally laden symbol structures are developed with the child becoming capable of learning to choose symbols at will. This enables the youngster to construct a world that is not primarily intellectual but tonal as well. The child does not just get meaning and feeling from the world. He literally gives meaning and feeling to the world he "creates." In this respect affect is a primary aspect of symbolic thought itself. It is the basis for how we create our own world in our own image. It is the medium through which all living becomes, not a logical construct, but a work of art.

Through this early interaction and the influence of parental and societal attitudes and injunctions, the child begins to construct a memory bank of experiences, feelings, imagery and thoughts from which he can draw upon at will. It is from the selective recall of these prior experiences drawn from the memory bank that early preferences, likes, dislikes and biases are established and conditioned.

These conditioned affective-cognitive responses then become the formative material for the development of many negative, rigid core beliefs and deep-seated attitudes which can often be translated into verbal statements such as: "I am mainly a body"; "I am defenseless"; "I am worthless"; "I am fragile"; "I am powerless"; "I am vulnerable"; "I am alone"; "I am a victim"; "The world is a threatening place"; "The world appears meaningless"; "Being alive is a hazard"; "Too many things happen unaccountably"; "I am not really loved"; "Everybody is for himself"; "I can never do what I want to do"; "I am always being punished"; "I am not wanted"; "I am not needed"; "Health, happiness and survival are factors determined by pure chance."

Chapter 29

Conclusion: Beyond Traditional Therapy

The traditional medical and optometric model which is currently supported by the majority of those in the health professions considers disease, structural changes and ocular defects as being caused mainly by internal and external forces acting on the body. An ever-increasing amount of research is being directed toward uncovering genetic, physiological, metabolic and biochemical factors that might predispose individuals to developing these conditions. Research is also being directed toward discovering those external conditions such as pollutants, irritants, radiation, bacteria and viruses that might be the activating agents. The recent interest in dietary and nutritional factors is also part of this search for answers. It is recognized that there is such an entity as a psychosomatic illness which has its origin in mental states, but more often emotional and personality problems are considered to be a reaction to one's own physical condition, the taking of medication, biochemical disturbances or simply from an inability to adjust to situations. Many psychologists and psychiatrists support this view and direct their efforts toward having those under their care learn how to cope better with whatever situations they meet up with in life and to develop more assurance and peace of mind.

The behavioral optometric approach does not totally disagree with this view but considers a whole group of common disorders as arising from the interaction of the individual with the environment. It sees disturbances in physiological function as being a result of stress brought on by such factors as reading for meaning, working too close to the task, poor lighting, poor posture, poor seating and working conditions, viewing computer terminals, too many distractions, etc. There is greater recognition of the interplay between functions and an awareness that psychological and perceptual factors are involved. The most emphasis however is placed on the action of externally induced stressors that require some sort of systemic readjustment such as increased accommodation or changes in accommodative-convergence function in order to cope.

The psycho-behavioral model is sensitive to all the exceptions to the rule: all those cases which do not seem to conform to the scenario hypothesized by the other models. It accepts the validity of viral, bacterial, biochemical, nutritional, genetic, traumatic and environmental factors as being related to many problems but leans to the theory that stress is not caused by stressors but by how one reacts to those stressors. And how one reacts is influenced by the perception of the event. We create the stress. The stress is us. The psycho-behavioral approach is a consciousness-oriented philosophy that emphasizes the primacy of mental states and the role beliefs play in the process of adaptation and its end result, the adaptive outcome.

Behavioral optometrists have always worked on some aspect of behavior through the use of lenses, prisms and visual training. When psychological states were seen to improve in conjunction with optometric therapy, it was considered to be an added benefit. Some have reasoned that this was a carry-over from the increased feelings of self-worth and self-confidence that came from improved performance. Others have sensed rightly that they might be tapping directly into mental states through their therapy programs. The hypothesis presented here is that although there is a bidirectional relationship between mental states and all aspects of behavior, the more important route is from mind (beliefs and attitudes) to behavior.

If this hypothesis is reasonable, then consideration should be given to working on mental states directly in conjunction with the more traditional approaches. Many successful clinicians have been doing this without realizing it. Their approach, conversation and the way instructions are given during a visual training program stimulate greater awareness on the part of the patients about their own role in relation to the problem they are attempting to resolve.

This final chapter, which in some respects is a continuation of chapter 17, is devoted to providing some suggestions for going beyond the traditional optometric behavioral therapeutic approach. This does not imply that these recommendations should be a substitute for conventional therapy nor that they must be followed in order to complete a case successfully. They are simply ideas that could help the reader personally and professionally. They might provide greater insight into how mental states influence perception, personality and performance. They might also help in achieving more positive results in resistant cases. Utilizing these procedures does not rule out the advantages of appropriate referrals for medical, allied medical, psychological or psychiatric help for the many who need it, but it does assume that there are many things people can do for themselves.

The suggestions aim at changing beliefs and attitudes that have been rigidly conditioned and deeply repressed. Therefore, they will not always be easy to do. There is always resistance to changing habits of any kind. Some have suggested that certain people could develop guilty feelings if they are told that they are in some way responsible for their condition. More often it has been found to be a liberating experience. It has provided hope when none was offered elsewhere. It has given people the feeling of being able to exert some form of control over their own lives with less dependence on outside sources. Even when it has been difficult to modify deep-seated core beliefs, many have achieved the feeling that life and living actually have some meaning after all. The following, therefore, is dedicated to a change of mind--a reorienting of life priorities and through it, to less stress and better vision.

Towards a Change of Mind

1. Assuming Responsibility

An important first step is to realize that everyone bears responsibility for the direction his or her life has taken and the adaptive outcomes that have come about. Each life is an individual act of creation, a work of art. Realize that the way a situation is perceived and reacted to is based on a "choice" that was made. The situation itself is neutral. Beliefs are not. Beliefs have tremendous power. Beliefs do not come out of thin air. They are selected, chosen and accepted. Since an inner voice may say otherwise, it is beneficial to keep repeating this thought. In time, it will begin to appear more reasonable.

Keep telling yourself that your problem, visual or otherwise, is a reflection of your attitudes, emotions and behavior--in that order. It did not simply happen by chance or inheritance. A negative approach or passive acceptance of the condition only reinforces the feeling of being a victim and of being vulnerable. Instead, reassure yourself that the condition is due to basic attitudes and beliefs and that these can be changed.

Consciously working to modify one's beliefs can be considered as the ultimate form of human creativity. Tell yourself that the beliefs you want to change are not true, that they are simply beliefs organized by the ego to sustain the feeling of "self." If you are not totally at peace with yourself, then be open to switching to new beliefs even as a game. In addition, keep telling yourself that your mind is more important than your body. It always was and always will be. Since the way you think, feel, act and see are under your control, take responsibility.

2. Making a Commitment

Do not merely accept the idea intellectually that deep-seated conditioned attitudes can be changed--believe it. This requires some degree of faith which is always served best when it is accompanied by determination, persistence and vigilance. Convince yourself that it can be done and will be done. Acknowledge the fact that your mental state is the most important part of your existence and that changing it will alter your perception, personality and performance. Then, make a commitment and work toward that goal with diligence. Any other way is simply idle self-talk and empty self-promises. Trust that the process will work. Create that state of mind where you know that you will be successful regardless of what others may say, what you yourself might have thought previously or what "objective" evidence seems to indicate. Want it above all else. A positive attitude is far more productive than a negative one.

3. Becoming More Aware

Become aware of the hidden attitudes and beliefs that have such a powerful effect on your perception and emotions and the way you respond to events. One way of accomplishing this is to observe how you interpret and react to given situations. Then reason backward to what beliefs or core attitudes might underlie that response. It might be helpful to write these beliefs down. A second method is to flow with the emotion that is experienced as a result of a given interaction and sense what thought may be behind it. You might discover that these thoughts involve expressions of guilt, vulnerability, being out of control, being unworthy, or the feeling of being a stranger in an alien existence. Do not get angry at the emotions, thoughts or the symptoms these foster. Simply accept them for what they are. Just be aware of them while reassuring yourself that you are fine the way you are and there is nothing wrong in having these feelings or thoughts. Let them run their course while you observe them passively. Fear or anger reinforce the negative emotions and make them stronger. Sometimes underlying thoughts surface more easily and with less resistance through the use of relaxation or meditative procedures which act to subdue distracting inner talk and images. In any case, try to become more aware of those key attitudes that mold your life. As Siegel (1984) puts it, "The healing process demands that patients know who they were so they know who to become."

4. Playing Positive Mind Games

Relaxation, meditation, exercise, visualization or visual training are often used as an end without realizing that they act mainly by setting the stage for change to occur. If one's attitudes and beliefs are rigidly maintained, only limited progress can be achieved from these activities. Changes will tend to be short lasting and will not fully generalize. One may feel better, see better and function better with less discomfort but unless core beliefs are changed in the process, progress will be temporary. The individual may be able to cope better in current situations but may not be able to do as well in future encounters.

To aid in changing basic beliefs, it was recommended that responsibility be assumed for "choosing" them in the first place, that one should trust that they can be changed and that one should become introspective in order to allow the beliefs to surface to conscious awareness from their repressed or suppressed state. The next step is to play a series of "positive thinking" verbal and visual mind games geared to changing these basic beliefs and reinforcing desired behavioral goals. These verbal fantasizing and visualization games should not be thought of as child's play. Believe that they will be effective, that they will work. Imagination happens to be one of the most powerful tools for developing and maximizing positive beliefs and lessening the effect of those beliefs that are negative and disabling. Accept the fact that pretending and fantasizing can truly alter outcomes. Also keep telling yourself that your present beliefs may have served a purpose when you were young and needed guidance and structure for your safety but there is less need to hang onto them when they are obviously no longer useful.

Regardless of how you feel, see or act, repeatedly tell yourself that you see better, feel better, look better and are better. Repeat Emile Coue's (1974) famous dictum: "Day by day, in every way, I am getting better and better." In using positive thinking, try not to reinforce negative states. Do not say, "I'm tired; I'm angry; I'm ill;" or "I don't see well." Say positive things to yourself such as, "I will relax and feel refreshed; I will be calm and happy; part of me is healthy, and I see well up close and will see better at distance." Then, in some small way, act as if you believe what you are telling yourself. Silva and Miele (1977) also recommended periodically saying, "Negative thoughts, negative suggestions, have no influence over me at any level of mind."

Use visualization and fantasize being in situations where you are seeing better, feeling better, looking better and acting better.

Play at trying to create or reproduce the emotion you want. If you are angry, try to simulate the emotion of joy. If you are tense, try to simulate feeling relaxed. If you are fearful, try to simulate being calm and feeling secure. If you feel separated from the world, try to simulate the feeling of being at one with the universe. If you are physically ill, pretend you are completely well. Regardless of how you feel, just play this game as often as possible. Avoid being a symptom watcher. Even when it is difficult, think positively, visualize positively and act positively. In this way you can prod your beliefs into new and more desirable directions. Keep in mind that unless the old beliefs are modified or replaced, the physical symbols they represent will tend to persist. If you find difficulty in imagining yourself being happy, content, fearless, etc., it may be that you inwardly feel that you do not deserve to have these positive feelings. Here, too, try to discover the hidden beliefs behind these feelings. As Rosanes-Berrett (1970) puts it, "It is essential for the implicit to become explicit....Genuine change occurs only when one is in touch with self. Then ears that dared not hear, eyes that dared not see, and a heart that dared not feel can come to life again and function normally."

In addition, whatever you do that triggers a more positive feeling or that makes you feel more relaxed, happier and peaceful should be pursued and repeated often. It could be an event, being with a certain person, doing a particular task, playing a particular game, listening to certain types of music, viewing art, dancing or singing. Doing this is both healthful and therapeutic. Similarly, avoid doing those activities that bring on or reinforce negative thoughts or emotions. As Harold Arlen's World War II song suggests, "Ac-cent-tchu-ate the positive. E- lim-in-ate the negative. Latch on to the affirmative. Don't mess with Mr. In-between."

Finally, to turn stress and visual stress around and enjoy life and living, these "positive thinking" techniques have to be used to eliminate fear. Fear stems from feeling inadequate and vulnerable. The feeling of inadequacy and vulnerability stems from interpretation, from comparing yourself to what you believe "should be." Fear arises when there is a conflict between what is actually experienced and what you think should be experienced. It is triggered by not perceiving and accepting things as they are. Therefore, perceiving and accepting people, things and events as they are without judgment, interpretation or expectation, without any "shoulds" becomes a guarantee of inner tranquillity.

To achieve this goal, "positive thinking" should also cultivate the feeling that everyone is all he needs to be and that no one is better or worse than any other. This includes visualizing that you and everyone else are part of something greater, all- encompassing and totally supportive. Depending on personal outlook, this quality can be called God, Universal Energy Source, Infinite Mind, Life Force, the Absolute, or the Ineffable. The name does not matter. The concept does. Those who feel this support are less fearful. Those who see themselves and everyone else as part of this greater universal quality are usually more accepting, less judgmental, more tolerant and more forgiving. Feeling protected, less competitive, less judgmental and less fearful is the ultimate goal of any effective stress-reduction regimen. To achieve this state requires accepting life for its own sake as simply a series of experiences and challenges to be enjoyed, flowing with events and letting go. Or, as was written to me by a friend and colleague, words appropriate enough to end this book, "Walk with God. Don't lead."

EPILOGUE

The real pleasure one gets out of science
is to see what everybody else has seen before,
but to think what no one else has even thought.
AlbertSzent-Gyorgyi

References

Abeloff, MD & LR Derogatis
1977 Psychologic aspects of the management of primary and metastatic breast cancer. Progres Clin & Biol Research 12:505- 516.
Achterberg, J & GF Lawlis
1978 Imagery of cancer. Champaign, IL: Instit for Personality and Ability Testing.
1975 A Course in Miracles. Tiburon, CA: Foundation for Inner Peace.
Adel, NI.
1963 Role of the ciliary musculature in accommodation and myopia. Opt J Rev Optom 100:23:23-26.
1966 Electromyographic and entopic studies suggesting a theory of action of the ciliary muscle in accommodation for near and its influence on the development of myopia. Am J Optom Arch Am Acad Optom (Jan) 43:27-38.
Ader, R.
1982 Psychoneuroimmunology. New York: Academic Press.
Alexander, FM.
1969 The resurrection of the body. New York: Dell Publ. Co.
1974 The Alexander technique, Ed. E Maisel. London: Thames & Hudson.
Allen, P.
1965 Social environmental factors in the etiology of functional esotropia. Am Orthoptic J 15:72-84.
Allwitt, LF.
1981 Two neural mechanisms related to modes of selective attention. J Exper Psychol: Human perception and performance 7:2:324-332.
Ambler, BA & DL Finklea
1976 The influence of selective attention in peripheral and foveal vision. Percept Psychophy 19:6:518-524.
Anderson, A.
1982 How the mind heals. Psychol Today (Dec) 16:12:51-56

Anderson, R & KE McClearey
1984 Pointing toward a poetic of personal growth. Revision (Spring) 7:1:35-48.
Andreassi, JL.
1980 Psychophysiology: human behavior and physiological response. Oxford: Oxford University Press.
Angle, J & DA Wissman
1978 Age, reading and myopia. Am J Optom Physiol Opt (May) 55:5 (May):302-308.
Antonovsky, A.
1974 Health, stress & coping. San Francisco, CA: Jossey- Bass.
Appel, RJ & JW Streff
1963 Optometric child vision care and guidance. Santa Ana, CA: Optom Extension Prog Curr II 7:4 & 5:13-22.
Appley, MH & R Trumbull
1967 On the concept of psychological stress, in Psychological stress, Ed. M Appley & R Trumbull. New York: Appleton-Century- Crofts.
Arehart-Treichel, J.
1980 Biotypes. New York: Time Books.
Assagioli, R.
1973 The act of will. New York: Viking.
1976 Psychosynthesis. Penguin.
Attinson, B.
1959 A preliminary report on the incidence of esophoria among mentally retarded patients. Am J Optom Arch Am Acad Optom (Feb) Monog:246.
Axelrod, J & TD Reisine
1984 Stress hormones: their interaction and regulation. (May 4) 224:452-459.
Bahnson, CB & MB Bahnson
1964 Denial and repression of primitive impulses and of disturbing emotions in patients with malignant neoplasm. In Psychosomatic aspects of neoplastic disease. Ed. DM Kissen and LL LeShan. Philadelphia, PA: Lippincott.
1969 Role of the ego defenses: denial and repression in the etiology of malignant neoplasm. Ann NY Acad Sci 125.
Balliet, R, A Clay & K Blood
1982 The training of visual acuity in myopia. J Am Optom Assoc 53:9:719-724.

Baldwin, WR.
1981 A review of statistical studies of relations between myopia and ethnic, behavioral and physiological characteristics. Am J Optm Physiol Opt (July) 58:7:516-527.

Barchas, JD, et al.
1978 Behavioral neurochemistry: neuroregulators and behavioral states. Science 200:964-973.

Bardiger, M & AL Stock
1972 The side effects of sucrose-containing diets low in protein on ocular refraction in the rat. Nutr Soc Proc 31.

Barker, S.
1978 The Alexander technique. New York: Bantam Books.

Bates, WH.
1920 The cure of imperfect sight by treatment without glasses. New York: Central Fixation Public. Co./Brooks Press.

Baum, A, JE Singer & CS Baum
1982 Stress and the environment, in Environ stress, Ed. GW Evans. Cambridge, England: Cambridge University Press.

Beatty, J.
1977 Activation and attention in the human brain. In The human brain, Ed. MC Wittrock, et al. Englewood Cliffs: NJ: Spectrum/Prentice Hall.

Beatty, J & D Kahneman
1966 Pupillary changes in two memory tasks. Psychonom Sci 5:371-372.

Beatty, J & GL Wagoner
1978 Pupillometric signs of brain activation vary with level of cognitive processing. Science (Mar 17) 199:1216-1218.

Beck, J & B Ambler
1972 The effects of concentrated and distributed attention on peripheral acuity. Percept Psychophys 12:33-38.

Becker, E.
1982 Growing up rugged: Fritz Perls and Gestalt therapy. ReVision (Fall) 5:2.

Beckwitt, ML.
1971 Stress and strabismus. Isr Ann Psychiatry (Apr) 9:1:11- 29.

Bedrossian, RH.
1966 Treatment of progressive myopia with atropine. Proceedings of the 20th Inter Cong Ophthal, Munich, Excerpta Medica.
1971 The effects of atropine on myopia. Ann Ophthal 3:8:891- 897.
Beedle, SL & FA Young
1976 Values, personality, physical characteristics and refractive error. Am J Optom Physiol Opt 53:735-739.
Belcastro, PA & RS Gold
1983 Teacher stress & burnout: implications for school health personnel, J of School Health (Sept) 53:7:404-407
Bell, GR.
1978 A review of the sclera and its role in myopia. J Am Optom Assoc (Dec) 49:12:1399-1403.
1980 The Coleman theory of accommodation and its relevance to myopia. J Am Optom Asso 51:582-588.
Benson, H.
1979 The relaxation response. New York: Wm. Morgan & Co.
Benson, H & W Proctor
1984 Beyond the relaxation response. New York: Times Books.
Berlyne, DE.
1960 Conflict, arousal and curiosity. New York: McGraw-Hill.
Berman, A.
1976 Visual and neuropsychological disturbance in emotional maladjustment. In Basic visual processes and learning disability, Ed. G Leisman. Springfield, IL: C.C.Thomas.
Bernard, C.
1859 Lecons sur les proprietes physiologiques et les alterations pathologiques des liquides d l'organisme. Paris: Bailliere.
1865 An introduction to the study of experimental medicine. Reprinted and transcribed (1957). New York: Dover.
1878 Lecons sur les phenomenes de la vie communs aux animaux et aux vegetaux. Paris: Bailliere.
Bieber, JC.
1974 Why nearpoint retinoscopy with children? Optom Wkly 65:54-57,78-82.

Birnbaum, MH.
1978 Holistic aspects of visual style: a hemispheric model with implications for vision therapy. J Am Optom Assoc (Oct) 49:10:1133-1141.
1979 Management of the low myopia pediatric patient. J Am Opt Assoc (Nov) 50:11:1281-1289.
1980 Nearpoint visual stress. Presented to the 10th annual COVD meeting (audio-cassette), Instatape, Pasadena, CA.
1981a Clinical management of myopia. Am J Optom Physiol Opt (July) 58:7:554-559.
1981b Symposium on perspectives on myopia: concluding remarks. Am J Optom Physiol Opt (July) 58:7:567-569.
1981c Esotropia, exotropia and cognitive/perceptual style. J Am Optom Assoc (Aug) 52:8:635-639.
1984a The power of visual training. J Am Optom Assoc (Apr) 55:4:257-260.
1984b Nearpoint visual stress: a physiological model. J Am Optom Assoc (Nov) 55:11:825-835.
1985a An esophoric shift associated with sustained attention. Am J Optom Physiol Opt 62:11:732-735.
1985b An esophoric shift associated with sustained attention. Invitational Skeffington symposium transcript (C. Croisant: Waldport, OR).

Bloom, FE.
1981 Neuropeptides. Sci Am 245:4:148-168.

Booth, G.
1969 General and organic--specific object relationships in cancer. Ann NY Academ Sci 164:568-577.

Borysenko, J.
1984 Psychoneuroimmunology: behavioral factors and the immune response. Revision (Spring) 7:1:56-65.

Bovard, EW.
1961 A concept of hypothalamic functioning. Perspec Biol Med 5:52-60.

Bowan, MD.
1984 Myopia, astigmia and hyperopia: stress diseases: an initial report on a unifying hypothesis. 29th annual invitational Skeffington symposium transcript (C. Croisant: Lebanon, OR).

Bresler, DE.
1984 Mind-controlled analgesia: the inner way to pain control. In Imagination and healing. Ed. AA Sheikh. Farmingdale, NY: Baywood Pub Co.

Bresler, DE & R Turbo
1979 Free yourself from pain. New York: Simon and Schuster.
Brody, JE.
1983 Heart attacks: turmoil beneath the calm. New York Times, June 21.
1984a Heart attacks and behavior; early signs are found. New York Times, Feb 14.
1984b Improve your health - change your type A behavior. Family Circle, May 29.
1984c Modifying type A behavior reduces heart attacks. New York Times, Aug 7.
Brown, BB.
1977 Stress and the art of biofeedback. New York: Harper & Row.
1980 Supermind. New York: Harper and Row.
Bruner, JS.
1966 The perfectability of intellect. In Knowledge among men. Ed. PH Oehser. New York: Simon & Schuster.
1983 In search of mind. New York: Harper & Row.
Bruner, JS, RR Olver & PM Greenfield, Eds.
1966 Studies in cognitive growth. New York: Wiley & Sons.
Burchfield, SR.
1979 The stress response: a new perspective. Psychosom Med 41:661-672.
Campbell, J.
1982a Grammatical man: information, entropy, language and life. New York: Simon & Shuster.
1982b Cancer: positive images. Brain/Mind Bull (June 21) 7:11:1-2.
Cannon, WB.
1928 Neural organization for emotional expression. In Feelings and emotions: The Wittenberg symposium, Ed. ML Reymert, Worcester, MA: Clark Univ. Press.
1929 Bodily changes in pain, hunger, fear and rage. New York: Appleton-Century-Crofts.
1932 The wisdom of the body. New York: Norton.
1935 Stresses and strains of homeostasis. Am J Med Sci 189:1-14.
Cannon, WB & D Paz
1911 Emotional stimulation of adrenal secretion. Am J Physiol 28:64-70.

Carrington, P.
1977 Freedom in meditation. New York: Doubleday.
Cheung, WY.
1982 Calmodulin. Sci Am 246:6.
Chvapil, M.
1967 Physiology of connective tissue. London: Butterworths.
Ciaranello, RD.
1983 Neurochemical aspects of stress. In Stress, coping, and development in children, Ed. N Garmezy and M Rutter. New York: McGraw-Hill.
Ciuffreda, KJ, et al.
1982 Use of eye movement auditory biofeedback in the control of nystagmus. Am J Optom Physiol Opt 59:5:396-409.
Cobb, S & RM Rose
1973 Hypertension, peptic ulcer and diabetes in air controllers. J Amer Med Assoc 224:489-492.
Cohen, L.
1961 The role of eye and neck proprioceptive mechanisms in body orientation and motor coordination. J Neurophysiol 24:1-11.
Cohen, S.
1980 Aftereffects of stress on human performance and social behavior: A review of research and theory. Psychol Bulletin 88:1:82-108.
1981 Sound effects on behavior. Psychol Today (Oct) 15:38- 49.
Cohen, SI, AJ Silverman, B Bressler & BM Schmavonia
1961 Problems in isolation studies, in sensory deprivation, Ed. P Solomon, et al. Cambridge, MA: Harvard Univ Press:114-129.
Cohn, H.
1886 The hygiene of the eye in schools. Trans. WP Trumbull. London: Simpkins, Marshall & Co.
Coleman, DJ.
1970 A unified model for the accommodative mechanism. Am J Ophthal 69:1063-1079.
Cornsweet, TN.
1970 Visual perception. New York: Academic Press.
Costello, M.
1974 The effect of stress and relaxation on autonomic function as measured by the resting state of accommodation. University Park, PA: Penn. State Univ.

Coue, E.
1974 Self-mastery through autosuggestion. New York: Samuel Weiser.
Coulter, HL.
1972 Homeopathic medicine. Washington, DC: Amer Inst for Homeopathy.
1973 Homeopathic influences in nineteenth century allopathic therapeutics. Washington, DC: Amer Inst for Homeopathy.
Criswell, MH & DA Goss
1983 Myopia development in non-human primates: a literature review. Am J Optom Physiol Opt 60:3:250-268.
Cross, AJ.
1911 Dynamic Skimetry in theory and practice. New York: AJ Cross Optical Co.
Curtin, BJ.
1970 Myopia, a review of its etiology, pathogenesis and treatment. Surv Ophthalmol 15:1-17.
1979 Physiologic vs. pathologic myopia: genetics vs. environment. Ophthalmology 86:681-691.
Darwin, C.
1872 The expression of emotion in man and animals. London: Murray.
Davson, H.
1976 The physiology of the eye. 3rd ed. New York: Academic Press.
DeBono, E.
1968 New think: the use of lateral thinking in the generation of new ideas. New York: Basic Books.
1969 Mechanism of mind. New York: Simon & Shuster.
Descartes, R.
1958 Discours de la method (1637). In Descartes philosophical writings, trans. NK Smith. New York: Random House.
Dilts, R, et al.
1980 Neuro-linguistic programming. Cupertino, CA: Meta Publ.
Dohrenwend, BS & BP Dohrenwend, Eds.
1974 Stressful life events: their nature & effects. New York: John Wiley & Sons.

Donders, FC.
1972 On the anomalies of accommodation and refraction of the eye. Rev. Ed. Boston: Milford House, (London: New Sydenham Society, 1864).
Dubos, R.
1965 Man adapting. New Haven, CT: Yale Univ Press.
1979 Mirage of health. New York: Harper & Row.
Dubos, R & JP Esconde
1979 Quest: reflections on medicine, science & humanity. New York: Harcourt, Brace, Jovanovich.
Duke-Elder, WS.
1946 Textbook of ophthalmology. Vol. 1. St. Louis, MO: CV Mosby Co.
Dunbar, F.
1947 Mind and body. New York: Random House.
Dunn, JP & S Cobb
1962 Frequency of peptic ulcer among executives, craftsmen and foreman. J Occup Med 4:343-348.
Easterbrook, JA.
1959 The effect of emotion on cue utilization and the organization of behavior. Psychol Rev 66:183-201.
Ebenholtz, SM
1983 Accommodative hysteresis: a precursor for induced myopia? Invest Ophthalmol Vis Sci 24:513-515.
Edinger, EF.
1972 Ego and archtype. New York: GP Putnam & Sons.
Eliot, RS.
1979 Stress and the major cardiovascular disorders. Mt. Kisco, NY: Futura Publ Co.
Eliot, RS & DL Breo
1984 Is it worth dying for? A self-assessment program to make stress work for you, not against you. New York: Bantam Books.
Eliot, RS & JC Buell
1983 The role of the CNS in cardiovascular disorders. Hospital Practice (May) 18:5:189-199.
Eliot, RS, et al.
1982 Bio-behavioral perspectives on coronary heart disease, hypertension and sudden cardiac death. Acta Med Scand (suppl.) 660:203-213.
Eliot, TS.
1963 Collected poems 1909-1962. New York: Harcourt Brace and World.

Elliott, GR & C Eisdorfer, Eds.
1982 Stress and human health: analysis and implications of research. New York: Springer Publishing Co.

Elul, R & PL Machiafava
1964 Accommodation of the eye as related to behavior in the cat. Arch Ital Biol 102:616-644.

Epstein, YM.
1982 Crowding stress and human behavior. In Environmental stress, Ed. GW Evans. Cambridge, England: Cambridge Univ Press.

Erickson, P.
1978 Crystalline-lens position and accommodative efficiency. Am J Optom Physiol Opt (Aug) 55:8:571-575.

Eriksen, B & J Rohrbaugh
1970 Some factors determining efficiency of selective attention. Am J Psychol 83:330-342.

Everly, Jr. GS & R Rosenfeld
1981 The nature and treatment of the stress response. New York: Plenum Press:131-141.

Eysenck, MW.
1982 Attention and arousal: cognition and performance. New York: Springer and Verlag.

Eysenck, SBG.
1956 An experimental study of psychogalvanic reflex responses of normal, neurotic and psychiatric subjects. J Psychosom Res 1:258-272.

Fauci, AS.
1975 Corticosteroids and circulating lymphocytes. Transplant Proc 7:37-40.

Fehmi, LH & G Fritz
1980 Open focus: the attentional foundation of health and well-being. Somatics (Spring):24-30.

Feldenkrais, M.
1972 Awareness through movement. New York: Harper & Row.

Feldman,, J.
1950 Myopia, vitamin A and calcium. Am J Ophthal 33:777-785.

Flax, N.
1984 A current look at the OEP B1 and B2 case typings. J Optom Vis Develop (Mar) 15:1:19-21.

Flom, MC, et al.
1980 Control of unsteady eccentric fixation in amblyopic eyes by auditory feedback of eye position. Invest Ophthal Vis Sci 19:11:1371-1381.

Forrest, EB.
1959-60a A modern philosophy of vision. Opt Wkly 50,51.
1960b A modern philosophy of vision. Part IV - The story of stress. Opt Wkly (Jan 28) 51:172.
1960c A modern philosophy of vision. Opt Wkly (Mar 31) 635- 636.
1960d A modern philosophy of vision. Opt Wkly, (May 5) 905- 907.
1967 Vision and the language arts--A study in hemispheric reciprocity. Opt J & Rev of Optom, 104:2.
1968-70 The psychobiology of visual behavior. Santa Ana, CA: Opt Extension Prog Curr II:41 & 42, Ser. 1 & 2.
1972 The visual/auditory-verbal syndrome. J Am Optom Assoc 43:8:873-883.
1976a Clinical manifestations of visual information processing, J Am Optom Assoc Part I:47:1 & Part II:47:1.
1976b Clinical manifestations of visual information processing, J Am Optom Assoc Part I (Jan) 47:1:73-80.
1976c Clinical manifestations of visual information processing, J Am Optom Assoc, Part II:47:4:499-507.
1978 Perspectives on visual function. In Vision: its impact on learning, Ed. R Wold. Seattle, WA: Special Child Publications:111-190.
1980a Stress: A re-definition. J Am Optom Assoc 51:6:600-604.
1980b Astigmatism as a function of visual scan, head scan and head posture. Am J Optom Physiol Opt (Nov) 57:11:844-860.
1981a A new model of functional astigmatism. J Am Optom Assoc (Nov) 52:11:889-897.
1981b Feedback and the visual process. J Am Optom Assoc (Sept) 52:9:717-724.
1981c Visual imagery: an optometric approach. Santa Ana, CA: Optom Extension Prog
1984a The innate vs. the learned: visual imagery and the role of experience. J Am Optom Assoc (Jan) 55:1:43-46.
1984b An operational approach to information processing strategies. J Am Optom Assoc (Feb) 55:2:103-107.
1984c Eye scan therapy for astigmatism. J Am Optom Assoc (Dec) 55:12:894-901.

1984d The tyranny of the premise: aspects of a psycho-behavioral philosophy of visual function. J Optom Vis Develop (Dec) 15:4:3-9.

Frankl, V.
1960 Paradoxical intention: a logotherapeutic technique. Am J Psycho 14:520-535.

Freedman, JL.
1975 Crowding and behavior. New York: The Viking Press.

French, JD.
1957 The reticular formation. Sci Am (May).

Freud, S.
1938 The basic writings of Sigmund Freud. New York: Modern Library.
1949 An outline of psychoanalysis. New York: Norton.

Friedman, E.
1981 Vision training program for myopia management. Am J Optom Physiol Opt (July) 58:7:546-553.

Friedman, N & RH Rosenman
1959 Association of specific overt behavior pattern with blood and cardiovascular findings. J Amer Med Assoc 169:1286- 1296.
1974 Type A behavior and your heart. New York: Alfred A. Knopf.

Friedman, M & D Ulmer
1984 Treating Your Type A behavior and your heart. New York: Alfred A. Knopf.

Fritz, G & L Fehmi
1982 The open focus handbook: the self-regulation of attention in biofeedback training and everyday activities. Princeton, NJ: Biofeedback Computers, Inc.

Fromm, E & DT Suzuki
1960 Zen buddhism and psychoanalysis. New York: Harper.

Garber, J & ME Seligman, Eds.
1980 Human helplessness: theory & applications. New York: Academic Press.

Gardiner, PA.
1958 Dietary treatment of myopia in children. Lancet 1:1152- 1155.

Gardner, R, J Beltramo & R Krinsky
1975 Pupillary changes during encoding, storage and retrieval of information. Percept Mot Skills 41:951-955.

Gardner, R, P Philip & S Radacy
1978 Pupillary changes during recall in children. J Exp Child Psychol 25:168-172.
Gawron, VJ.
1981 Differences among myopes, emmetropes and hyperopes. Am J Optom Physiol Opt 58:9:753-780.
1983 Ocular accommodation, personality and autonomic balance. Am J Optom Physiol Opt (July) 60:7:630-639.
Gazzaniga, M.
1967 The split brain in man. Sci Am (Jan).
Gazzaniga, M & JE LeDoux
1978 The integrated mind. New York: Plenum Press.
Gellhorn, P.
1953 Physiological foundations of neurology and psychiatry. Minneapolis, MN: Univ of Minn. Press.
Gesell, A, F Ilg, G Bullis & GN Getman
1949 Vision, its development in infant and child. New York: Paul B. Hoeber.
Gesell, A, et al.
1950 Vision, its development in infant and child. New York: Harper & Bros.
Getman, GN.
1960 Techniques and diagnostic criteria for the optometric care of children's vision. Santa Ana, CA: Optom Extension Prog
1977a Optometric analysis of visual performance. Santa Ana, CA: Opt Extension Prog Curr II (Nov) 50:2:2,9.
1977b Optometric analysis of visual performance. Santa Ana, CA: Opt Extension Prog Curr II (Dec) 50:2:3:14-15.
1978a Optometric analysis of visual performance. Santa Ana, CA: Optom Extension Prog Curr II (Jan) 50:2:4:19.
1978b Optometric analysis of visual performance. Santa Ana, CA: Optom Extension Prog Curr II (July) 50:2:10:61.
1978c Optometric analysis of visual performance. Santa Ana, CA: Optom Extension Prog Curr II (Sept) 50:2:12:72.
Getman, GN & NC Kephart
1958 Book retinoscopy. Santa Ana, CA: Optom Extension Prog. Curr II 23:2:10,11:65-74.
Gillilan, LA.
1954 Clinical aspects of the autonomic nervous system. Boston, MA: Little, Brown & Co.

Gimbel, HV.
1973 The control of myopia with atropine. Can J Ophthalmol 8:4:527-532.
Glass, DC.
1977 Behavior patterns, stress & coronary disease. Hillsdale, NJ: Lawrence Erlbaum Assoc. Wiley & Sons.
1980 Behavioral medicine: an emergent field (as quoted by Constance Holden) Science (July 25) 479-481.
Glass, DC & JE Singer
1972 Urban stress: experiments on noise and social stressors. New York: Academic Press.
Goldrich, SG.
1983 Oculomotor biofeedback for exotropia. Am J Optom Physiol Opt 59:4:306-317.
Goldstein, K.
1939 The organism. New York: American Book Co.
1963 Human nature in the light of psychopathology. New York: Schocken Books.
Grace, W & D Graham
1952 Relationship of special attitudes and emotions to certain bodily diseases. Psychosom Med 14:243-251.
Granit, R.
1962 The visual pathway. In the eye: Vol. 2, Ed. H Davson. New York: Academic Press.
Gray, J.
1971 The psychology of fear and stress. New York: McGraw- Hill.
Green, E & A Green
1977 Beyond biofeedback. New York: Delacorte/Lawrence.
Green, EE, AM Green & ED Walters
1970 Voluntary control of internal states: psychological and physiological, Part I. J Transpers Psychol 2:1-26.
Greene, M.
1970 Submarine myopia in the minuteman launch control facility. J Am Optom Assoc (Dec) 41:12:1012-1016.
Greene, PR.
1980 Mechanical considerations in myopia: relative effects of accommodation, convergence, intraocular pressure, and the extraocular muscles. Am J Optom Physiol Opt (Dec) 57:12:902-914.

Greenspan, SB.
1974 M.E.M. retinoscopy. The refraction letter. Rochester, NY: (Dec) S Bausch & Lomb.
1981 Research studies of bifocals for myopia. Am J Optom Physiol Opt (July) 58:7:536-540.
Greer, S.
1979 Psychological attributes of women with breast cancer. Cancer Det & Prev 2:289-294.
Greer, S & T Morris
1975 Psychological attributes of women who develop breast cancer: a controlled study. J Psychosom Res 19:147-153.
Greer, S, T Morris & KW Pettingale
1979 Psychological response to breast cancer: effect on outcome. Lancet 13:785-787.
Groffman, S.
1978 Psychological aspects of strabismus and amblyopia--a review of the literature. J Am Optom Assoc (Sept) 49:9:995-999.
Grossfield, HD.
1951 Visual space and physical space. J Psychol 32:25-33.
Gruber, E.
1978 The treatment of myopia with atropine: a clinical study. Proceedings of the 23rd Inter Congr Ophthal. Amsterdam, Excerpta Medica.
Guilford, JP.
1967 The nature of human intelligence. New York: McGraw- Hill.
Gurin, J.
1979 Chemical feelings. Science 80 (Nov-Dec) 28:33.
Hahnemann, S.
1961 Organon of medicine, 6th ed. Trans. RE Dudgeon. Calcutta: Roy Publ House.
Hall, H.
1984 Imagery and cancer. In Imagination and healing, Ed. AA Sheikh. Farmingdale, NY: Baywood Publ.
Harmon, DB.
1958 Notes on a dynamic theory of vision, 3rd Ed. Austin, TX: Research Publications.
1966a The rationale in developmental vision training. Seminar transcript (March).
1966b Restrained performance as a contributing cause of visual problems. Optom Wkly (July 7).
1966c Seminar transcript (Nov).

1969 Orientation and localization. Seminar Notes (Sept).
Hart, HH.
1949 The eye in symbol or symptom. Psychoanaly Rev 36:1-2.
Harvey, J.
1978 Diaphragmatic breathing: a practical technique for breath control. The Behavior Therapist 1:13-14.
1983 Understanding perception. Dawn 3:2:16-24.
Hasker, L & RT Zacks
1979 Automatic and effortful processes in memory. J Exper Psychol Gen 108:3:356-388.
Hasset, J.
1978 Teaching yourself to relax. Psychol Today. (Aug) 12:3.
Haynes, HM.
1960 Clinical observations with dynamic retinoscopy. Optom Wkly 51:2243-46, 2306-09.
Haynes, SG, M Feinleib & WB Kannel
1980 The relationship of psychosocial factors to coronary heart disease. In the Framingham Study III. Eight-year incidence of coronary heart disease. Am J Epidemiol 111:37-58.
Heaton, JM.
1968 The eye: phenomenology and psychology of function and disorder. London: Favistock.
Hebb, DO.
1961 Organization of behavior. New York: John Wiley & Sons.
Henderson, T.
1934 The constitutional factor in myopia. Trans Ophthalmol Soc UK:54:451-454.
Hendrickson, H.
1980 The behavioral approach to lens prescribing. Santa Ana, CA: Opt Extension Prog.
Henry, JP & D Ely
1976 Biologic correlates of psychosomatic illness. In Biological foundations of psychiatry. Eds. R Grenell & S Galay. New York: Raven Press.
Hensley, DE.
1984 Overcoming career burnout. Optom Manag (Feb) 20:2:75- 79.
Herrick, CJ.
1924 Neurological foundations of animal behavior. New York: Holt.

Hess, EH.
1965 Attitude and pupil size. Sci Am.
1972 Pupillometrics: a method of studying mental, emotional and sensory processes. In Handbook of psychophysiology, Eds. NS Greenfield and RA Sternbach. New York: Holt, Rinehart and Winston: 491-531.
1975a The tell-tale eye. New York: Van Nostrand Rheinhold.
1975b The role of pupil size in communication. Sci Am (Nov).
Hess EH & JM Polt
1964 Pupil size in relation to mental activity during simple problem solving. Science 143: 1190-1192.
Hildreth, HR, et al.
1947 The effect of visual training on existing myopia. Am J Ophthalmol 30:1563-1576.
Hirons, R, et al.
1978 Biofeedback treatment of strabismus: case studies. J Am Opt Assoc 49:8:875-882.
Hoffman, JW, et al.
1982 Reduced sympathetic nervous system responsivity associated with the relaxation response. Science (Jan) 215:190- 192.
Holden, C.
1978 Cancer and the mind: how are they connected? Science (June 23) 200:1363-1369.
Holmes, DS, et al.
1984 Task-related arousal of Type A and Type B persons: level of challenge and response specificity. J Pers Soc Psychol 46:6:1322-1327.
Holmes, TH & RH Rahe
1979 The social readjustment scale. J Psychoso Research 11:213-218.
Hoskins, RG.
1946 The biology of schizophrenia. New York: Norton.
Hughes, A.
1972 Vergence in the cat. Vision Res 12:1961-1994.
Humphreys, MS & W Revelle
1984 Personality, motivation and performance: a theory of the relationship between individual differences and information processing. Psychol Rev 91:2:153-184.
Hynes, E.
1956 Refractive changes in normal young men. Arch Ophthal (Nov) 56:5:761-767.

Jacobs, S, et al.
1970 Life stress and respiratory illness. Psychosom Med 32:233-242.
Jacobson, E.
1929 Progressive relaxation. Chicago: Univ of Chicago Press.
1957 You must relax. New York: McGraw-Hill.
1964 Anxiety and tension control. Philadelphia, PA: Lippincott.
James, W.
1948 Psychology. (1890) Reprinted. Cleveland, OH: World Publishing Co. Living Library Edition.
1950 Principles of psychology. New York: Dover.
Janisse, MP.
1977 Pupillometry: the psychology of the pupillary response. Washington, DC: Hemisphere Publi. Co.
Jaynes, J.
1976 The origin of consciousness in the breakdown of the bicameral mind. Boston, MA: Houghton-Mifflin.
Jemmott, JB, III & SE Locke
1984 Psychosocial factors, immunologic mediation and human susceptibility to infectious diseases: how much do we know? Psychol Bull 95:1:78-108.
Jenkins, CD.
1971 Psychologic and social precursors of coronary disease. N Eng J Med 284:244-255,307-317.
1976 Recent evidence supporting psychologic and social risk factors for coronary disease. N Eng J Med 294:987-994.
Jenkins, CD, RH Rosenman & SJ Zyzanski
1974 Prediction of clinical coronary heart disease by a test for the coronary-prone behavior pattern. N Eng J Med 23:1271- 1275.
Johnson, CA.
1976 Effects of luminance and stimulus distance on accommodation and visual resolution. J Optom Soc Am 66:138-142.
Jung, CG.
1923 Psychological types. New York: Harcourt.
1953-58 Collected works. Ed. Bollingen. New York: Pantheon.
1976 Psychological types - collected works, Vol. 6. Ed. RFC Hull. Princeton, NJ: Princeton Univ. Press.

Kabat-Zinn, J.
1982 An outpatient program in behavioral medicine for chronic pain patients based on the practice of mindfulness meditation: theoretical considerations and preliminary results. Gen Hosp Psych 4:33-47.
1984 Coping with chronic pain: an interview with John Kabat- Zinn. ReVision (Spring) 7:1:66-72.
Kagan, J.
1983 Stress and coping in early development. In Stress, coping and development in children. Eds. N Garmezy & M Rutter. New York: McGraw-Hill.
Kagan, J & W Phillips
1964 Measurements of identification: a methodological note. J Abnorm Soc Psychol 69:442-444.
Kagan, J & BL Rosman
1964 Cardiac and respiratory correlates of attention and an analytic attitude. J Exp Child Psychol 1:50-63.
Kahneman, D.
1973 Attention and effort. Englewood Cliffs, NJ.: Prentice- Hall.
Kahneman, D & J Beatty
1966 Pupil diameter and load in memory. Science 154:1583- 1585.
Kaplan, JR, et al.
1983 Social stress and atherosclerosis in normocholesterolemic monkeys. Science (May 13) 220:733-735.
Kaplan, MM.
1965 A new theory of dynamic visual resolution and corneal accommodation. Optom Wkly (Sept 16,23,30; Oct 7,14,21,28; Nov 4,18; Dec 2).
1978-79 Vertical yoked prisms. Santa Ana, CA: Optom Extension Prog Curr II 51:1.
Katona, PG & F Jih
1975 Respiratory sinus arrhythmia: noninvasive measure of parasympathetic cardiac control. J App Physiol 39:801-805.
Keller, S, et al.
1981 Suppression of immunity by stress: effect of a graded series of stressors on nymphocyte stimulation in the rat. Science 213:1397-1400
Kelley, CR.
1971 New techniques in vision improvement. Los Angeles, CA: Interscience Research Institute.

Kelly, T, C Chatfield & G Tustin
1975 Clinical assessment of the arrest of myopia. Br J Ophthalmol 59:10:529-538.
King, S.
1981 Imagineering for health. Wheaton, IL: Quest/Theosophical Publ. House.
Kinney, J, et al.
1980 The vision of submariners and national guardsmen: logitudinal study. Am J Optom Physiol Opt (Aug) 57:8:469-478.
Kinsbourne, M.
1974 Mechanisms of hemispheric interaction in man. In Hemispheric disconnection and cerebral function. Eds. M Kinsbourne & WL Smith. Springfield, IL: Charles C Thomas.
Kissen, DM.
1963 Personality characteristics in males conducive to lung cancer. Br J Med Psychol 34:27-36.
1964 Relationship between lung cancer, cigarette smoking, inhalation and personality. Br J Med Psychol 37:203-216.
1967 Psychosocial factors, personality and lung cancer in men aged 55-64. Br J Med Psychol 40:29-43.
Kissen, DM & HJ Eysenck
1962 Personality in male lung cancer patients. J Psychosom Res 6:123-237.
Kissen, DM & LGA Rao
1969 Steroid excretion patterns and personality in lung cancer. Ann NY Acad Sci 128:476-482.
Knapp, AA.
1939 Vitamin D complex in progressive myopia: etiology, pathology and treatment: preliminary study. Am J Ophthal 22:1329- 1337.
Koffka, K.
1922 Gestalt psychology. Psychol Bulletin 19
Kohler, W.
1947 Gestalt psychology. New York: Liveright Publishing Corp.
Kolata, G.
1982 New theory of hormones proposed. Science 215:1383-1384.
Konrad, KW & MH Bagshaw
1970 Effect of novel stimuli on cats reared in a restricted environment. J Comp Physiol Psychol 70:157-164.

Koppel, GD.
1979-80 Nutrition and vision. Santa Ana, CA: Optom Extension Prog Curr II 52:1.
1983-84 Vision and nutrition. Santa Ana, CA: Optom Extension Prog Curr II 56:2.
Koppman, JW, RD McDonald & M Runzel
1974 Voluntary regulation of temporal artery diameter by migraine patients. Headache (Oct) 14:3:133-138.
Kraskin, RA.
1965 Stress-point retinoscopy. J Am Optom Assoc 36:5:416-419.
1981-83 Lens power in action. Santa Ana, CA: Optom Extension Prog Curr II, 54 & 55: 1 & 2.
Krieger, DT.
1983 Brain peptides: what, where and why? Science 222:975- 985.
Krishnamurti, J.
1968 Talks and dialogues. New York: Discus/Avon.
1982 The network of thought. New York: Harper & Row.
Kruger, PB.
1977 Brightness changes during book retinoscopy. Am Jr Optom & Physiol Optics 54:10:673-677.
1980 The effect of cognitive demand on accommodation. Am Jr Optom & Physiol Optics 57:7:440-445.
Ku, DN & PR Greene
1981 Scleral creep in vitro resulting from cyclic pressure pulses: applications to myopia. Am J Optom Physiol Opt (July) 58:7:528-535.
Lacy, JI.
1967 Somatic response patterning and stress: some revision of activation theory. In Psychological stress, Eds. MH Appley & R Trumbull. New York: Appleton-Century-Crofts.
Lancaster, WB.
1923 Certain anatomical and physiological considerations bearing on heterophoria. South Med J 16.
Lane, BC.
1981 Lead, vanadium and chromium affect eye focusing. Abstracts of the 109th Annual Meeting (Am Publ Health Assoc): 245-246.
1982 Myopia prevention and reversal: new data confirms the interaction of accommodative stress and deficit-inducing nutrition. J Inter Acad Prev Med (Nov) 7:3:17-30.

Lanyon, RI & JW Giddings
1974 Psychological approaches to myopia: a review. Am J Optom Physiol Opt 51:271-281.
Larabee, PE & FR Jones
1980 Behavioral effects of low plus lenses. Percept Mot Skills 51:913-914.
Laudenslager, ML, et al.
1983 Coping and immunosuppression: inescapable but not escapable shock suppresses lymphocyte proliferation. Science (Aug 5) 221:568-570.
Lazarus, RS.
1969 Patterns of adjustment and human effectiveness. New York: McGraw-Hill.
1981 Little hassles can be hazardous to health. Psychol Today (July) 15:7:58-62.
Leibowitz, HW.
1977 Visual perception and stress. In physical work and effort: proceedings of the international symposium, ED. G Borg. Wenner-Gren Center, NY: Pergamon.
Leibowitz, HS & DA Owens
1975 Anomalous myopias and the intermediate dark focus of accommodation. Science (Aug 22) 189:646-648.
1978 New evidence for the intermediate position of relaxed accommodation. Doc Ophthalmol 46:133.
LeShan, L.
1959 Psychological states as factors in the development of malignant disease: a critical review. J of Natl Cancer Instit 22:1-18.
1966 An emotional life history pattern associated with neoplastic disease. Ann NY Acad Sci 125:780-793.
1974 How to meditate. Boston, MA: Little, Brown & Co.
1977 You can fight for your life: emotional factors in the causation of cancer. New York: M Evans.
1982 The mechanic and the gardener: making the most of the holistic revolution in medicine. New York: Holt, Rinehart & Winston.
LeShan, L & M Reznikoff
1960 Psychological factors apparently associated with neoplastic disease. J Abnor & Soc Psychol 60:439-440.
LeShan, L & RE Worthington
1956 Personality as a factor in the pathogenesis of cancer: a review of the literature. Br J Med Psychol 29:49-56.

Levi, L.
1972 Introduction: psychosocial stimuli, psychophysiological reactions and disease. In Stress and distress in response to psychosocial stimuli. Ed. L. Levi. Oxford: Pergamon Press.

Levine, S.
1971 Stress and behavior. Sci Am 224:1:26-31.

Lewis, M & CD Wilson
1970 The cardiac response to a perceptual-cognitive task in the young child. Psychophysiology 6:411-420

Light, KC.
1981 Cardiovascular responses to effortful active coping. Psychophysiology 18:216-225.

Light, KC, et al.
1983 Psychological stress induces sodium and fluid retention in men at high risk for hypertension. Science (Apr 22) 220:429- 431.

Lilly, J.
1972 The center of the cyclone. An autobiography of inner space. New York: Julian Press.

Lindemann, H.
1973 Relieve tension the autogenic way. New York: Peter H Wyden Inc.

Lion, EG, C O'Neill & RB Prager
1943 Strabismus and children's personality reactions. Am J Orthopsychol 13:1:121-124.

Lipton, EL.
1970 A study of the psychological effects of strabismus. Psychoanaly. Study Child 25:146-174.

Locke, J.
1965 An essay concerning human understanding. Ed. M Cranston. New York: Colliers Books/Macmillan Co.

Lorenz, K.
1977 Behind the mirror. London: Methuen & Co.

Lovasik, JV & R Beauchamp
1982a Ocular accommodation in the cat after electrical stimulation of mesencephalic reticular formation. Am Jr Optom & Physiol Optics 59:9:726-734.
1982b Ocular accommodation neurons in brain stem of the alert cat. Am Jr Optom & Physiol Optics 59:10:785-794.

Lovgren, GK.
1977 The art of inner seeing. Sun City, AZ: Karl Bern Publ.

Lowenstein, O & IE Loenfeld
1950 Role of sympathetic and parasympathetic systems in reflex dilation of the pupil. Arch Neur & Psychia 64:313-340.
Lowman, CL.
1918 The effect of faulty skeletal alignment upon the eyes. Am J Orthop Surgery (Dec):459-492.
Ludlum, WM
1968 Applied myopia control. San Diego behavioral vision seminar, San Diego, CA: Transcript: 39-92.
1973 The visual evoked response. Presented to the third annual COVD meeting (audio-cassette), Pasadena, CA, Instatape.
Luria, SM, H Newmark & H Beatty
1970 Effect of a submarine patrol on visual processes. Report No. 641, US Navy Bureau of Medicine & Surgery, Navy Dept. Research Work Unit M4305.08-300 ID.06.
Luthe, W.
1963 Autogenic training: method, research and application in medicine. Am J Psychoth 17:174-195.
Luttgens, K & KF Wells
1982 Kinesiology: scientific basis of human motion, 7th Ed. Philadelphia, PA: Saunders College Publ.
Lynn, R.
1966 Attention, arousal and the orientation reaction. Oxford: Pergamon Press.
Macdonald, LW.
1962-65 Visual training. Santa Ana, CA: Optom Extension Prog Curr II 1-3.
1962 Visual training. Santa Ana, CA: Optom Extension Prog Curr II (Nov) 35:1:2:8.
1965 Visual training. Santa Ana, CA: Optom Extension Prog Curr II (Sept) 37:3:12.
Maddox, EE.
1893 The clinical use of prisms; and the decentering of lenses, 2nd ed. Bristol, England: John Wright & Sons.
Malmstrom, RV & RJ Randle
1976 Effects of visual imagery on the accommodation process. Percept Psychophys 19:5:450-453.
1984 Effect of a concurrent counting task on dynamic visual accommodation. Am Jr Optom & Physio Optics 61:9:590-594.

Malmstrom, RV, RJ Randle, JS Bendix & RJ Weber
1980 The visual accommodation response during concurrent mental activity. Percept Psychophys 28:5:440-448.
Maranto, G.
1974 The mind within the brain. Discovery (May) 5:5:34-43.
Marg, E.
1952 Flashes of clear vision and negative accommodation with reference to the Bates method of visual training. Am J Optom Arch Am Acad Optom 29:4:167-184.
Margach, CB.
1983 Dark loci of identification and centering. Santa Ana, CA: Optom Extension Prog Curr II (June): Literature and research review 8:9:58-63.
1984 Case typings and dark loci. Santa Ana, CA: Optom Extension Prog Curr II (July): Literature and research review 9:10:47-51.
Martin, B.
1961 The assessment of anxiety by physiological behavioral measures. Psychol Bull 58:234-255.
Maslach, C.
1976 Burned out. Hum Behav 5:16-22.
1982 Burnout: the cost of caring. Englewood Cliffs, NJ: Prentice-Hall.
Mason, JW.
1971 A re-evaluation of the concept of "non-specificity" in stress theory. J Psych Res 8:323-333.
1975 Confusion and controversy in the stress field. Part I & Part II. J Human Stress.
Marx, JL.
1979 New information about the development of the autonomic nervous system. Science 206:434-437.
Matthews, KA.
1982 Psychological perspectives on the Type A behavior pattern. Psychol Bull 91:2:293-323.
Maupin, EW.
1969 On meditation. In Altered states of consciousness, Ed. CT Tart. New York: John Wiley & Sons.
McGuinnes, D & K Pribram
1980 The neuropsychology of attention: emotional and motivational controls. In The brain and psychology, Ed. MC Wittrock. New York: Academic Press.

McKim, RH.
1980 Thinking visually. Belmont, CA: Lifetime Learning/Wadsworth.
McQuade, W & A Aikman
1974 Stress. New York: EP Dutton.
McQuerter, G.
1978 Cancer: clues in the mind. Sci News (Jan 21) 113:3:44- 45.
Mecacci, L & B Pasquali
1980 Eyewink, evoked potentials and visual attention. Percep Mot Skills 51:891-895.
Meldman, MJ.
1970 Diseases of attention and perception. Oxford: Pergamon Press.
Mershon, DH & TL Amerson
1980 Stability of measures of the dark focus of accommodation. Invest Ophthalmol Vis Sci 19:217-221.
Metchenbaum, D.
1975 A self-instructional approach to stress management: a proposal for stress inoculation training. In Stress and anxiety, Vol. 2, C Spielberger & I Sarason, Eds. New York: Wiley.
Meyer, C.
1981 Five words that can change your life. Let's Live (Nov).
Miele, J.
1985 Hyperopia reduction. Presentation to the Eastern States Optometric Congress, Ellenville, NY, April 29.
Miller, JA.
1982 Cells of babel. Science News 122:25 & 26.
Miller, JA & I Peterson
1984 Life in a maddening crowd. Science News, June 16.
Miller, RJ.
1978 Temporal stability of the dark focus of accommodation. Am J Optom Physiol Opt 55:447.
Mills, L.
1919 The effect of faulty craniospinal form and alignment upon the eyes. Am J Ophthal (Sept): 493-499.
Morgan, MW.
1960 The relationship of refractive error to bookishness and androgyny. Am J Optom Arch Am Acad Optom 37:171-185.

Morrow, L.
1981 The burnout of almost everyone. Time (Sept 21) 118(12):84.
Morse, DR & ML Furst
1979 Stress for success. New York: Van Nostrand Reinhold.
Morse-Peckham, R.
1926 The modern treatment of binocular imbalances. New York: Shuron Standard Opt Co.
1939 Some points in the correction of myopia, 5th ed. Clinical research report No. 1. Detroit, MI: The Optom Research Inst.
Moutinho, H & J Dos Santos
1968 Psychopedagogic aspects of observations and treatment of strabismic and amblyopic patients. In International strabismus symposium, A. Arruga, Ed. Giessen, Aug 9-12, 1966, Basel, S Karger, 1968:431-434.
Mucha, TF & RF Reinhardt
1970 Conversion reactions in student aviators. Am J Psychiatry (Oct) 127:4:493-497.
Mull, HK.
1948 Myopia and introversion. Am J Psychol 61:588-591.
Naranjo, C & R Ornstein
1971 On the psychology of meditation. New York: Viking.
Nathanson, JA & P Greengard
1977 Second messengers in the brain. Sci Am (Aug) 237:108- 119.
Neisser, U.
1967 Cognitive psychology. New York: Appleton-Crofts.
1976 Cognition and reality. San Francisco, CA: WH Freeman.
Nichols, AS.
1943a The pointer method of observing and training visual skills, Part I--a premise. Optom Wkly (Apr 29 & May 6) 359-361, 387-389.
1943b The pointer method of observing and training visual skills, Part II--the pointer method. Optom Wkly (May 13 & 20): 415-417, 443-444.
1943c The pointer method. J Am Optom Assoc (Dec) 15:5, 139- 144.

1945a Visual training. 5th Midwest Conference on Visual Training, St. Louis, MO, Transcript (July): 86-91.
1945b Conjecture on space perception and myopia. Optom Wkly (Sept 20) 36:33:927-929.
1946a Mirror technique of vision training - Part I. Optom Wkly (Sept 12) 37:37:1267-1269.
1946b Mirror technique of vision training - Part II. Optom Wkly (Sept 19) 37:38:1307-1310.
1947 1946 Year Book. Stockton, CA.
1948 Seminar Transcript. Pittsburgh, PA: (Apr 11-13).
1950 The principles of perceptual science. Seminar Transcript, Boston, MA: (Oct 25).
1951 The principles of perceptual science Seminar Transcript, Boston, MA: (May 26-30).

Novaco, RW.
1979 The cognitive regulation of anger and stress. In cognitive-behavioral interventions: theory, research and procedures, Eds. PC Kendall & SD Hollon. New York: Academic Press.

Nuernberger, P.
1981 Freedom from stress. Honesdale, PA: Himalayan Inter Instit.
1982 Breath, the way of balance. Dawn 2:2:18-23.

Oakley, KH & FA Young
1975 Biofocal control of myopia. Am J Optom Physiol Opt (Nov) 52:11:758-764.

Olmstead, JMD.
1944 The role of the autonomic nervous system in accommodation for far and near vision. J Nerv Men Dis 11:282- 290.

Ornstein, RE.
1972 The psychology of consciousness. San Francisco, CA: WH Freeman.
1977 The psychology of consciousness, 2nd Ed. New York: Harcourt Brace Jovanovich.

Owens, DA.
1979 The Mandelbaum effect: evidence for an accommodative bias toward intermediate viewing distances. J Optom Soc Am 69:646-652.

Owens, DA & HW Leibowitz
1983 Perceptual and motor consequences of tonic vergence. In Vergence eye movements: basic and clinical

aspects. Eds. CM Schor & KJ Cuiffreda. Boston, MA: Butterworths.

Owens, RL & KE Higgins

1983 Long-term stability of the dark focus of accommodation. Am J Optom Physiol Opt 60:1:32-38.

Page, J.

1984 Rural stress. Science 84 (Mar) 5:2:104-105.

Parker, MW.

1958 Protective-corrective program for young myopes. Optom Wkly 49:15:681-683.

Pelletier, KR.

1977 Mind as healer, mind as slayer. New York: Delacorte/Lawrence.

Penfield, W.

1975 The mystery of the mind: a critical study of consciousness and the human brain. Princeton, NJ: Princeton Univ Press.

Perls, FS.

1969 Ego, hunger and aggression. New York: Vintage Books/Random House.

Pert, A.

1981 The body's own tranquilizers. Psychol Today (Sept).

Peter, LC

1941 The extra-ocular muscles, 3rd Ed. Philadelphia, PA: Lea Febiger.

Pheiffer, CH.

1955 Book retinoscopy, J Am Optom Assoc 32:540-545.

Piaget, J.

1952 The origins of intelligence in children. New York: International Universities Press.

1955 Language and thought of the child. Cleveland, OH: World Publishing.

1971 Biology and knowledge. Chicago, IL: Univ of Chicago Press.

Pierce, JR.

1966-67 Research on the relationship between nearpoint lenses, human performance and physiological activity of the body. Santa Ana, CA: Optom Extension Prog Curr II Research Reports and Special Articles 39:1-12.

Pollie, DM, AJ Hafner & JH Krasnoff

1964 The strabismic child and his psychological adjustment. J Pediatric Ophthal (Apr) 1:2:60-63.

Pomeranz, RB.
1964 The use of bell retinoscopy to determine the nearpoint prescription. Optic J & Rev Optom 101:23:21-24.
Popper, K & J Eccles
1977 The self and its brain. Berlin: Springer International.
Pribram, KH.
1971 Languages of the brain. Monterey, CA: Brooks/Cole.
1981 Emotions. In Handbook of clinical neuropsychology, Eds. SK Filskov & TJ Boll. New York: Wiley/Interscience.
Pribram, KH & D McGuiness
1975 Arousal, activation and effort in the control of attention. Psychol Rev 82:2:116-149.
1978 Central control systems in attention and memory. Presented to the 8th Annual College of Optometrists in Vision Development Meeting, audiocassette. Instatape: Pasadena, CA.
Pugh, M.
1934 Classification of cases of concomitant strabismus based on the aetological factor. Br J Ophthal 18:446.
Radler, DH.
1956 Can nervous tension change structure of the eyes? Santa Ana, CA: Optom Extension Prog News, June 1956, Reprinted from Nov 6, 1955 issue of Indianapolis Star Magazine.
Rappaport, E.
1959 Anger, apathy and strabismus. Eye, Ear, Nose & Throat Mag (June) 38:6:473-482.
Reich, W.
1949 Character analysis, 3rd Ed. New York: Orgone Inst. Press.
Rice, TB.
1930a Physical defects in character, I-farsightedness. Hygeia 8:536-538.
1930b Physical defects in character, II-nearsightedness. Hygeia 8:646.
Roberts, WL & RD Banford
1967 Evaluation of bifocal correction technique in juvenile myopia. Optom Wkly 58:38:25-31; 58:39:21-30;58:40:23-28; 58:41:27-34; 58:43:19-26.

Rodin, J, S Solomon & J Metcalf
1978 Role of control in mediating perceptions of density. J of Person Soc Psychol 36:9:989-999.
Rosanes, MB.
1966 Psychological correlates of myopia compared to hyperopia and emmetropia. Doctoral dissertation, New York: Yeshiva Univ.
1967 Psychological correlates of myopia compared to hyperopia and emmetropia. J Proj Tech Person Asses 31:31-35.
Rosanes-Berrett, MB.
1970 Gestalt therapy as an adjunct treatment for some visual problems. In Gestalt therapy now, Eds. J Fagan & IL Shepherd. New York: Harper Colophon Books.
Rose, RM.
1980 Endocrine responses to stressful psychological events. Psych Clin of N Amer 2:53-71.
Rosenman, RH.
1978 Role of Type A behavior pattern in the pathogenesis of ischemic heart disease and modification for prevention. Adv Cardiol 25:35-46.
Rosenman, RH, et al.
1975 Coronary heart disease in the western collaborative group study: final follow-up experience of 8 1/2 years. J Am Med Assoc 233:872-877.
Rosenthal, DN, Ed.
1971 Materialism and the mind-body problem. Englewood Cliffs, NJ: Prentice-Hall.
Rotman, D.
1974 Molecular pathology of axial myopia: the role of ascorbic acid. Phil J Ophthal 6:2:71-74.
Rouse, MW, R London & DC Allen
1982 An evaluation of the monocular estimate method of dynamic retinoscopy. Am J Optom Physiol Opt 59:234-239.
Rouse, MW, RF Hutter & R Shiftlett
1984 Anormative study of the accommodative lag in elementary school children. Am J Optom Physiol Opt 61:11:693-697.
Rubin, LS.
1974 The utilization of pupillometry in the differential diagnoses and treatment of psychotic and behavioral disorders. In Pupillary dynamics and behavior. Ed. MP Janisse. New York: Plenum Press.

Rutter, M.
1981 Stress, coping and development: some issues and some questions. J Child Psychol Psychiat 22:4:323-356.
Ryle, G.
1949 The concept of mind. London: Hutchinsons Univ Library.
Saltysiak, BA.
1969 Getman retinoscopy. J Am Optom Assoc 2:1:14-17.
Sato, T.
1957 The causes and prevention of acquired myopia. Tokyo: Kanehara Shuppan Co.
Schapero, M & MJ Hirsch
1952 The relationship of refractive errors andGuilford-Martin temperament test scores. Am J Optom Arch Am Acad Optom 29:32-36.
Schmale, A & H Iker
1966 The psychological setting of uterine cervical cancer. Ann NY Acad Sci 125:807-813.
Schmeck, HM.
1984 Explosion of data on brain cell reveals its great complexity. Sci Times, New York Times (March 6).
Schor, CM.
1983 Fixation disparity and vergence adaptation. In Vergence eye movements: basic and clinical aspects. Eds. CM Schor & RJ Cuiffreda. Boston, MA: Butterworths.
Schor, CM, CA Johnson & RB Post
1984 Adaptation of tonic accommodation. Ophthal Physiol Opt 4:2:133-137.
Schultz, JH & W Luthe
1959 Autogenic therapy: a psychophysiologicalapproachin psychotherapy. New York: Grune & Stratton.
Schwartz, I & N Sandburg
1954 The effect of time in submarine service on vision. Report No. 253, US Navy Bureau of Medicine and Surgery (Aug 30).
Seligman, ME.
1975 Helplessness: on depression, development and death. San Francisco, CA: WH Freeman.
Selye, H.
1970 On just being sick. Nutrition Today 5:1.
1974a Stress and distress. Physician's World, New York: (March): 25-28.

1974b Stress without distress. New York: JB Lippincott.
1976 The stress of life, (1956) Rev. ed. New York: McGraw- Hill.
1978 They all looked sick to me. Hum Nature:58-63.
1979 Stress, cancer and the mind. In Cancer, stress and death. Eds. J Tache, H Selye & SB Day. New York: Plenum:11-19.

Septon, RD.
1984 Myopia among optometry students. Am J Optom Physiol Opt 61:12:745-751.

Serban, G.
1982 The tyranny of magical thinking. New York: EP Dutton.

Shaffer, M.
1982 Life after stress. New York: Plenum Press.

Shannon, CE.
1948 A mathematical theory of information. Bell Telephone Tech J 27:379-423, 623-656.
1951 Prediction and entropy of printed english. Bell Syst Tech J 30:1:50-64.
1956 The bandwagon. Transactions on informational theory IT 2:3 Institute of Electrical and Electronics Engineers.

Shavit, Y, et al.
1984 Opioid peptides mediate the suppressive effect of stress on natural killer cell cytoxicity. Science (Jan 13) 223:188-190.

Sherrington, A.
1947 Integrative action of the nervous system (2nd ed. preface). Cambridge, MA: Cambridge Univ Press.

Sherrod, DR.
1974 Crowding, perceived control and behavioral aftereffects. J App Soc Psychol 4:171-186.

Shipman, VI.
1954 A constriction of the perceptual field under stress. Paper presented to the Eastern Psychol Assoc, Philadelphia, PA.

Shotwell, AJ.
1984 Plus lens, prism and bifocal effects on myopia progressionin military students, Part II. Am J Opt Physiol Opt 61:2:112-117.

Shute, CCD.
1979 The McCollough effect. Cambridge, England: Cambridge University Press.

Siegel, B.
1984 The health of the healing profession: an interview with Bernard Siegel. ReVision (Spring) 7:1:87-93.
1968 Silence is Leaden. Sci News (July 6) 94.
Silva, J & P Miele
1977 The Silva mind control method. New York: Simon & Schuster.
Silverstein, S.
1981 A light in the attic. New York: Harper & Row.
Silvestri, R.
1978 CT: the astounding new confrontation therapy. New York: William Morrow & Co.
Simonton, OC & S Simonton
1975 Belief systems and management of the emotional aspects of malignancy. J Trans Psychol 7:1:29-48.
1978 Getting well again. New York: JB Tarcher.
Singer, RM.
1978 The relationship between usual refractive error and interests, values, personality, academic performance, and reading. Unpublished doctoral dissertation, Washington State Univ.
Skeffington, AM.
1931 Differential diagnosis in ocular examination. Chicago, IL: AJ Cox & Co.
1937 Transcript of lectures to the Oregon graduate clinic, Dec 1935. C. Croisant, Ed. Lectures of Portland and northwestern congresses 1935-36, Ft. Worth, TX: Text Publ. Co.
1938 Lectures to the Northwestern Congress 1937, Transcript Ed. C. Croisant, Ft. Worth, TX: Text Publishing Co.
1950a Nearpoint optometry. Santa Ana, CA: Opt ExtensionProg Curr II (Oct) 15:6:1.
1950b Nearpoint optometry. Santa Ana, CA: Opt ExtensionProg Curr II (Nov) 6:2:8.
1950c Nearpoint optometry. Santa Ana, CA: Opt ExtensionProg Curr II (Dec) 6:3:15.
1950d Practical applied optometry. Santa Ana, CA: Optom Extension Prog Curr II (Sept) 22:9:81.
1950e Practical applied optometry. Santa Ana, CA: Optom Extension Prog Curr II (Nov) 22:11:97-98.
1950f Nearpoint optometry. Santa Ana, CA: Opt Extension Prog Curr II 15:6:1.

1951a Nearpoint optometry. Santa Ana, CA: Opt Extension Prog Curr II (June) 6:9:44.
1951b Nearpoint optometry. Santa Ana, CA: Opt Extension Prog Curr II (July) 6:10:47-49.
1951c Nearpoint optometry. Santa Ana, CA: Opt Extension Prog Curr II (Sept) 6:12:63.
1953 Northwest Congress of Optometry, Transcript, Portland, OR: The Reporter Co. (Dec 13-15).
1955a Practical applied optometry. Santa Ana, CA:Opt Extension Prog Curr II (May) 27:5.
1955b Northwest Congress of Optometry, Transcript, Portland, OR: The Reporter Co. (Dec 11-13).
1957a Practical applied optometry. Santa Ana, CA:Opt Extension Prog Curr II (Jan) 29:4:31.
1957b Practical applied optometry. Santa Ana, CA: Opt Extension Prog Curr II (Nov) 30:2:10.
1957c Northwest Congress of Optometry, Transcript, Portland, OR: The Reporter Co. (Dec 15-17).
1957d Practical applied optometry. Santa Ana, CA: Opt Extension Prog Curr II (June) 29:9:77-78.
1957e Practical applied optometry. Santa Ana, CA: Opt Extension Prog Curr II (July) 29:10:85-88.
1957f Practical applied optometry. Santa Ana,CA: Opt Extension Prog Curr II (Oct) 30:1:2.
1958a Practical applied optometry.Santa Ana, CA: Opt Extension Prog Curr II 30:9.
1958b A dynamic approach to vision. Seminar Notes.
1958c Northwest Congress of Optometry, Transcript, Portland, OR: The Reporter Co. (Dec 14-16).
1961 Northwest Congress of Optometry, Transcript, Portland, OR: The Reporter Co. (Dec 3-5).
1963 The time factor in lens fitting. In Clinical optometry in theory and practice. Santa Ana, CA: Opt Extension Prog Curr II (Nov) 36:2.
1964 Introduction to clinical optometry. Santa Ana, CA: Opt Extension Prog Curr II (Nov) 1:2:9-10.
1965a Clinical optometry in theory and practice. Santa Ana, CA: Opt Extension Prog Curr II (Jan) 37:4:40.
1965b Clinical optometry in theory and practice. Santa Ana, CA: Opt Extension Prog Curr II (Feb) 37:5:48.
1965c Clinical optometry in theory and practice. Santa Ana, CA: Opt Extension Prog Curr II (Apr) 37:7:60.
1966 Clinical optometry in theory and practice. Santa Ana, CA: Opt Extension Prog Curr II (Feb) 38:5:46.

1969 Northwest Congress of Optometry, Transcript. Portland, OR: The Reporter Co. (Dec 7-9).
1970 Northwest Congress of Optometry, Transcript. Portland, OR: The Reporter Co. (Dec 6-8).
1972 The Skeffington Colloquium Outline. Santa Ana, CA. Opt Extension Prog.

Sklar, LS & H Anisman
1979 Stress and coping factors influence tumor growth. Science (Aug 3) 205:513-515.
1981 Stress and cancer. Psychol Bull 89:3:369-406.

Snyder, R. Ed.
1980 Buckminster Fuller: an autobiographical monologue/scenario. New York: St. Martins Press.

Sokolov, EN.
1960 Neuronal models and the orienting reflex. In The central nervous system and behavior. Ed. MAB Brazier. New York: Josiah Macy, Jr. Foundation.

Sorsby, A.
1934 The pre-myopic state: its bearing on the incidence of myopia. Trans Ophthalmol Soc UK:54:459-467.

Sperry, R.
1964 The great cerebral commissure. Sci Am (Jan).

Stevens, DA & HH Wolff
1965 The relationship of myopia to performance on a test of leveling-sharpening. Percept Mot Skills 21:399-403.

Streff, JW.
1962 Preliminary observations on a non-malingering syndrome. Opt Wkly 53:12:536-537.
1969a A visual syndrome-a review. Eastern Seaboard-Visual Training Conference. Transcript C Croisant, Lebanon, OR:84-88.
1969b Juvenile bilateral functional amblyopia: the Streff syndrome, monograph (in preparation).
1977 The Cheshire study: changes in incidence of myopia following program intervention. In Frontiers in visual science, proceedings of the Univ of Houston College of Optom dedication symposium, Eds. SJ Cool & EL Smith. New York: Springer-Verlag.

Streff, JW & RJ Apell
1960 Gesell institute of child development, seminar notes, New Haven, CT.

Stroebel, CF.
1982 QR: the quieting reflex. New York: Berkeley Books.
Sutton, AA.
1975 Application of lenses and prisms behaviorally. Presented to the 5th Annual COVD Meeting (audio-cassette). Instatape: Monrovia, CA.
1985 Optometric analysis of visual space. In Building a visual space world. Santa Ana, CA: Optom Extension Prog Curr II (Jan) 57:1:4:23.
Swogger, G, Jr.
1981 Toward understanding stress: a map of the territory. J of School Health (Jan): 29-33.
Tarler-Benlolo, L.
1978 The role of relaxation in biofeedback training: a critical review of the literature. Psychol Bull 85:4:727-753.
Tart, CT, Ed.
1975 Transpersonal psychologies. New York: Harper & Row.
Thomas, CB, KR Duszynski & JW Shaffer
1979 Family attitudes reported in youth as potential predictors of cancer. Psychosom Med 41:287-302.
Thomas, CB & R Greenstreet
1973 Psychobiological characteristics in youth as predictors of five disease states: suicide, mental illness, hypertension, coronary heart disease and tumor. Johns Hopkins Med J 132:16-43.
Thompson, SC.
1981 Will it hurt less if I can control it? A complex answer to a simple question. Psychol Bull 90:1:89-101.
Thorington, J.
1947 Methods of refraction. Philadelphia, PA: Blackiston Co.
Thorington, RN.
1900 Refraction and how to refract, 2nd Ed. Philadelphia, PA: P Bladiston & Son.
Trachtman, JN.
1978 Biofeedback of accommodation to reduce functional myopia: a case report. Am J Optom Physiol Opt (June) 55:6:400- 406.
1984 Biofeedback of accommodation to reduce myopia--a seven-year follow-up. Abstracts from the 1984 Annual Meeting Am Acad Optom, Am J Optom Physiol Opt (Oct) 61:10:93.

Trachtman, JN, V Giambalvo & J Feldman
1981 Biofeedback of accommodation to reduce functional myopia. Biofeedback Self Regul 6:547-564.
Tramonti, J.
1984 Notes from the lectures of Dr. Ann Sutton-Nichols. Northeast Vision Conference Transcript, Northampton, CT: (Jan 18).
1984 Transcendental Truth. Omni (Jan) 6:4:129.
Trevarthen, CB.
1968 Two mechanisms of vision in primates. Psychologische Forschung 31:299-337.
Trotter, RJ.
1973 The biological depths of loneliness. Science News (Mar 3) 103:9:140-142.
1983 Baby face. Psychol Today (Aug) 17:8:14-20.
Tryon, WW.
1975 Pupillometry: asurveyofsourcesofvariation. Psychophysiology 12:90-93.
Tubesing, DA.
1982 Kicking your stress habits. Duluth, MN: Whole-Person Assoc., Inc.
Tucker, DM.
1981 Lateral brain function, emotion and conceptualization. Psychol Bull 89:19-46.
Tucker, DM & PA Williamson
1984 A symmetric neural control systems in human self-regulation. Psychol Rev 91:2:185-215.
Vale, W. et al.
1981 Characterization of a 41-residue ovine hypothalmic peptide that stimulates secretion of corticotropin and beta- endorphin. Science (Sept) 213:1394-1397.
Vanacea, P.
1971 New concept in the treatment of progressive myopia. Br J Ophthal 3:1105-1108.
Van Alphen, GW.
1952 A comparative psychological investigation in myopes and emmetropes. Proc K Nederl Akad Wet 55:5:689-696.
Van Brocklin, MD, et al
1981 Biofeedback enhanced strabismus therapy. J Am Optom Assoc 52:9:731-736.
Vander, AJ.
1981 Nutrition, stress and toxic chemicals: an approach-to environment-health controversies. Ann Arbor, MI: Univ. of Michigan Press.

Veninga, RL & JP Spradley
1981 The work stress connection: how to cope with-job burnout. Boston, MA: Little, Brown & Co.
Vithoulkas, G.
1971 Homeopathy: medicine for the new man. New-York: Avon Books.
Von Neumann, J.
1966 Lectureat Univ. of IL, 1949. In Theory of self-reproducing automata. Ed. AW Burks, Urbana, IL: Univ of Illinois Press.
Wallace, RK & H Benson
1972 The physiology of meditation. Sci Am 226:2:84-90.
Wallman, JH, J Turkel & J Trachtman
1978 Extreme myopia produced by modest change in early visual experience. Science (Sept 29) 201:1249-1251.
Ware, J.
1813 Observations relative to near and distant sight of different persons, Part I. Philosophical transactions of the royal society. 31-50.
Weiner, N.
1949 Cybernetics. New York: John Wiley & Sons.
Weisz, CL.
1980 The accommodative resting state: theoretical implications of lens prescriptions. Rev Optom Part I (July) 117:7:60-70; Part II (Aug) 117:8:62-74.
Wertheimer, M.
1921 Untersuchungen Zur Lehr von der Gestalt, I - Psychol. Forsch 1
1923 Untersuchungen Zur Lehr von der Gestalt, II - Psychol. Forsch 4.
1944 Gestalt theory. Social Research 11.
Westheimer, G.
1957 Accommodative measurements in empty fields. J OptSoc Am 47:714-718.
Westheimer, G & SM Blair
1973 Accommodation of the eye during sleep and anesthesia. Vision Res 13:1035-1040.
Whatmore, GB & DR Kohli
1979 Dysponesis: a neurophysiologic factor in functional disorders. In Mind/body integration. Eds. E Pepper, etal. New York: Plenum Press.
Williams, Jr., RB.
1984 An untrusting heart. The Sciences (Sept/Oct).

Williams, Jr., RB, et al.
1982 Type A behavior and elevated physiological and neuroendocrine responses to cognitive tasks. Science (Oct 29) 218:483-485.
Wingerson, L.
1984 Some ways to break the tension of work. Newsday (April 30).
Wittkower, E & H Warners, Eds.
1977 Psychosomatic medicine. New York: Harper & Row.
Wolf, KS.
1984 The role of accommodation and binocular vergence in visual fatigue induced by near work at a video display terminal or hard copy. Santa Ana, CA: Optom Extension Prog. In Research reports and special articles. Curr II (Dec) 57:3:61-76.
Wood, DJ.
1927 Calcium deficiency in blood with reference to spring catarrh & myopia. Br J Ophthal 11:394-401.
Woods, AC.
1946 Report from the Wilmet Institute on the results obtained in the treatment of myopia by visual training. Am J Ophthal 29:28-57.
Woolf, D.
1964 Visual function in theory and practice. Santa Ana, CA: Optom Extension Prog Curr II (June) 36:1:9:44.
Yamamoti, K.
1979 Children's ratings of the stressfulness of experiences. Develop Psychol 15:581-582.
Young, FA.
1955 Myopes versus non-myopes--a comparison. Am J Optom Arch Am Acad Optom 32:180-191.
1961 The development and retention of myopia by monkeys. Am J Optom Arch Am Acad Optom (Oct) 38:10:545-555.
1963a The effect of restricted visual space on the refractive error of the young monkey eye. Invest Ophthalmol Vis Sci 2:6:571- 577.
1963b Reading measures of intelligence and refractive errors. Am J Optom Arch Am Acad Optom 40:5:257-264.
1965a The etiology of myopia. Optom Wkly (July 15): 17-24.

1965b The effect of atropine in the development of myopiain monkeys. Am J Optom Arch Am Acad Optom (Aug) 42:8:439-449.

1967a Myopia and personality. Am J Optom Arch Am Acad Optom 44:3:192-201.

1967b Present state of knowledge regarding the mechanisms giving rise to refractive anomalies. Refractive anomalies of the eye. NINDB monograph No. 5, Washington, DC: US Govt Printing Office: Public Health Service Publ. No. 1687.

1969 The transmission of refractive errors within Eskimo families. Am J Optom Arch Am Acad Optom 46:9:676-685.

1970 The transmission of refractive errors within Eskimo families--a reply to Arnold Sorsby. In "Letters to the editor," Am J Optom Arch Am Acad Optom (Mar) 47:3.

1975 The development and control of myopia in human and subhuman primates. Contacto (Nov) 16-31.

1977 The nature and control of myopia. J Am Optom-Assoc (April) 48:4:451-457.

1981 Primate myopia. Am J Optom Physiol Opt 58:560-566.

Young, FA, RJ Beattie, FJ Newby & MT Swindal
1954 The Pullman study: a survey of Pullman school children. Am J Optom Arch Am Acad Optom 31:111-121, 192-203.

Young, FA, GA Leary & DN Farrer
1971 Four years of animal studies of chimpanzee vision. Am J Optom Arch Am Acad Optom (May) 48:5:407-416.

Young, FA, RM Singer & D Foster
1975 The psychological differentiation of the male myopes and non-myopes. Am J Optom Physiol Opt 52:10:679-686.

Zeiger, CA.
1976 A psychological approach to the improvement of-myopia. Doctoral dissertation, Boulder Univ of Colorado.

Zubek, JP & L Wilgosh
1963 Prolonged immobilization of the body: changes in performance and in the electroencephalogram. Science 140:306-308.

Zuckerman, M.
1964 Perceptual isolation as a stress situation: a review. Am Med Arch Gen Psychiat 2:255-276.